CW00525188

ISBN 978-0-332-68195-5
PIBN 10987827

SMITHSONIAN INSTITUT

UNITED STATES NATIONAL.

RECENT FORAMINIFER.

A DESCRIPTIVE CATALOGUE OF SPECIMENS I THE U. S. FISH COMMISSION STEAMER ALE

BY

JAMES M. FLINT, M. D., U. S. N.,

Honorary Curator, Division of Medicine, U. S. National Museum.

From the Report of the U. S. National Museum for 1897, pages 249-3(with eighty plates.

WASHINGTON:

GOVERNMENT PRINTING OFFICE.

1899.

RECENT FORAMINIFERA.

A DESCRIPTIVE CATALOGUE OF SPECIMENS DREDGED BY THE U. S. FISH COMMISSION STEAMER ALBATROSS.

BY

JAMES M. FLINT, M. D., U. S. N.,
Honorary Curator, Division of Medicine, U. S. National Museum.

RECENT FORAMINIFERA. A DESCRIPTIVE CATALOGUE OF SPECIMENS DREDGED BY THE U. S. FISH COMMISSION STEAMER ALBATROSS.

By James M. Flint, M. D., U. S. N.,

Honorary Curator, Division of Medicine, U. S. National Museum.

PREFACE.

The purpose of this catalogue is to record the results of an examination of a portion of the bottom material obtained during the dredging operations of the U. S. Fish Commission steamer *Albatross*, and at the same time to furnish a convenient book of reference for those who are, or may become, sufficiently interested to continue the study of this material.

The examination, while very far from exhaustive, has been pursued with greater or less diligence, as time and opportunity offered, for several years. Material from about one hundred and twenty-five stations has been carefully studied, and specimens from more than a hundred localities have been preserved and identified. Of these localities, fifty-eight are in the North Atlantic Ocean, twenty-one in the Gulf of Mexico, seven in the Caribbean Sea, one in the South Pacific, and five in the North Pacific. The depths at these stations vary from 7 to 2,512 fathoms.

The figures in illustration are from photographs of mounted specimens on exhibition in the U. S. National Museum, Division of Marine Invertebrates. A uniform enlargement of about 15 diameters has been maintained in the figures, sometimes at a sacrifice of detail in the smaller specimens which would have been made clearer by the use of a higher magnifying power, but for the purpose of identification it is believed to be more useful to mark distinctly the relative size of the objects. The exhibition series has been mounted expressly for public display. The individuals of each species are attached in various attitudes to the bottom of the shallow cavity of a concave, blackened disk of brass. For security each disk is provided with a removable fenestrated brass cap having a top of thin glass. These disks are arranged in concentric rows upon a large circular metal plate which occupies the place of the stage of an ordinary microscope. The circular plate is given both a rotary and a to-and-fro movement by means of a friction roller and a rack and pinion, so that all the mounts may be successively brought under the microscope. The specimens thus arranged are inclosed in a box having a glass top, through which the objective of a microscope projects.[1]

In the following catalogue the classification of Mr. H. B. Brady has been followed, as presented in "The report on the Foraminifera collected by H. M. S. *Challenger*," and his definitions of families and genera have been appropriated bodily. The analytical table is also compiled chiefly from the above-mentioned report. The descriptions of species have been prepared after study of the reserve series as well as of the typical specimens reproduced in the illustrations.

The localities given are only those from which specimens have been taken in selecting the series exhibited and in reserve, and do not profess to represent the distribution of the species.

A supplementary table gives the latitude, longitude, and depth of water of the stations referred to in the catalogue.

THE FORAMINIFERA.

The Foraminifera are minute aquatic, mostly marine, animals, having semifluid bodies, composed of granular protoplasm, inclosed in shells or "tests" either secreted by the animal or built up of available foreign material, such as mud, sand, sponge spicules, or dead shells. In zoological classification they belong to the Rhizopod group of the Protozoa, and are distinguished from other members of the group by the single character of the reticulated form assumed by their pseudopodia when extended.

These minute animals are interesting objects of study, geologically and biologically as well as esthetically. As objects of beauty they arrest the attention of even the casual observer by the delicacy of their structure as well as the symmetry and variety of their forms. Geologically they are of interest because they are among the most ancient and abundant of fossils and also the most efficient of rock builders. Biologically they are instructive examples of the powers and possibilities of an individualized bit of protoplasm—"a little particle of apparently homogeneous jelly, changing itself into a greater variety of forms than the fabled Proteus, laying hold of its food without members, swallowing it without a mouth, digesting it without a stomach, appropriating its nutritious material without absorbent vessels or a circulating system, moving from place to place without muscles, feeling (if it has any power to do so) without nerves, propagating itself without genital apparatus, and not only this, but forming shelly coverings of a symmetry and complexity not surpassed by those of any testaceous animals."[1]

From the resemblance of some of the shells of the foraminifera to those of the nautilus, they were for a long time regarded as minute cephalopod mollusks; that is, among the highest of the invertebrates, and it was not until the year 1835 that their true nature was discovered and announced by M. Dujardin to the French Academy of Sciences. Since that time the study of this order of animals has been pursued by able naturalists, and the results of their investigations appear in a voluminous literature. Much yet remains to be learned of the life history of the animal, but its zoological position is established and its importance in the economy of nature recognized.

As fossils the foraminifera are common in all geological systems from the Devonian upward, but they are especially abundant in Mesozoic and Cenozoic time. The chalk and many of the most extensive limestone beds are formed principally of their remains. As to present habitat, their shells are found wherever dredgings are made, all over the ocean floor except in the polar regions. A few species are "pelagic;" that is, they are found living at or near the surface of the water, but the weight of evidence is in favor of the conclusion that the vast majority of them pass all stages of life at the bottom, where they are found. In the experience of the naturalists of the *Albatross* it was rare to find any but the most minute and thin-shelled forms in the surface dredgings, and still more rare for any to be taken in the "wing nets" that were usually attached to the dredging apparatus.

The living foraminifer is a minute bit of viscid, granular protoplasm, without organs or tissues, without differentiation of substance into outer membrane and inner contents, and in most instances without evident nucleus or contractile vesicle. A nucleus has been recognized in a few individuals, and hence this characteristic element of most living cells is inferred to be present in all the members of the order. Like other Rhizopods, it has the power to protrude any parts of its body as "pseudopodia," for the purpose of locomotion or the prehension and absorption of food. It differs, however, from the other Rhizopods in that the pseudopodia do not necessarily remain distinct, but flow together whenever they touch one another, forming sometimes an elaborate and extended network of protoplasmic threads, which, however, may be readily retracted and flow again into the body mass, leaving no indication of their previous existence.

How the function of nutrition is accomplished and the nature and condition of the organic material used as food by these minute animals is not yet determined. Without doubt the pseudopodia are capable of seizing and inclosing small organic particles with which they may come in contact, and any part of the protoplasmic body, of which the pseudopodia are but temporary extensions, is able to digest and assimilate the

[1] Carpenter, Introduction to the Study of the Foraminifera.

nutritive portion. To further account for the necessary food supply, it is believed that the Foraminifera absorb organic matter held in solution by the sea water. This theory is the more easily accepted since we know that they have the power to separate inorganic matter (carbonate of lime in particular) from its solution, with which to construct, wholly or in part, their shells.

Of the process of reproduction little is known beyond the fact of multiplication by gemmation and fission. Every part of this simple animal being sufficient unto itself for purposes of nutrition and growth, it follows that a fragment of the protoplasmic body cast off from the parent becomes at once a new individual and the possible founder of a fresh colony. But it is not in accord with what we know of the life histories of other living things that this process of subdivision or propagation by cuttings or shoots can go on indefinitely. It is more likely that some kind of sexual reproduction takes place, the manner of which is yet to be demonstrated.

The most striking characteristic of this simple, semifluid animal, of indefinite and changeable shape, is its ability to construct a shell or test of definite form in which to shelter itself. This shell or test may be irregular, simple, and rude in construction, or symmetrical and of great delicacy and beauty, in variety of forms rivaling the shells of the Mollusca, of which it was long thought to be a diminutive example.

Structurally there are three quite definite and distinct types of testaceous covering. The first, to begin with the lowest and least common, is the "chitinous" test—a thin, transparent, yellowish or brownish membranous investment secreted by the animal. It has one or more general apertures, but is not perforated with fine foramina, and there is no means of communication between the inside and outside of the test except by the general apertures. The foraminifera with this kind of shell have been grouped in the single family of *Gromidæ*. As a rule they inhabit only fresh or brackish water. They have not been found in deep-water marine collections, and do not appear in the following catalogue.

The second type is the so-called "arenaceous" test. This is an investment constructed of grains of sand, or of the dead shells of other foraminifera, or of sponge spicules, or even of mud, cemented together more or less firmly by means of a calcareous cement secreted by the animal. Usually it has one or more general apertures of comparatively large size, and in addition there may be minute orifices between the sand grains, or other substances of which the test is constructed, through which the delicate threads of protoplasm can be projected. The surfaces may be rough and coarse or smooth and highly finished.

The third type of structure is a true shell, composed almost entirely of carbonates of lime and magnesia separated by the animal from their solution in the sea water, and fixed in solid form. It is through the agency of the foraminifera principally that the limestone, which is constantly being dissolved by rains and carried by rivers to the sea, is restored to the solid crust of the earth. Of these "calcareous" shells there are two kinds, quite distinct in appearance, known as "porcellanous" and "hyaline." The former are usually white, opaque, shining with the peculiar luster of porcelain, and "imperforate;" the latter are transparent, glassy, and "perforate," more or less densely, by minute, parallel, unbranched tubes for the passage of delicate pseudopodia. In both kinds there are usually one or more comparatively large, general apertures. Surface marking, or "ornamentation," is common in both the porcellanous and hyaline shells. In the former they take the form of striations or pittings, more or less regular and conspicuous; in the latter, of ridges, tubercles, or spines, of clear nontubular shell-substance, varying constantly in number and prominence among individuals of the same species.

Architecturally the first and most obvious division of these shells is into single-chambered (monothalamous, or unilocular) and many-chambered (polythalamous, or multilocular). While the primitive form of both the single and many chambered shells is evidently globular, yet the possibilities of ultimate conformation, depending chiefly upon direction of growth, are very great. Thus a monothalamous shell, beginning as an incomplete spherical chamber, may become ovate, flask-shaped, spindle shaped, star-shaped, or tubular, and the tubular form may be straight, curved, coiled, or quite irregular. And these forms pass from one into another by quite insensible degrees. The polythalamous shell is a consequence of the process of reproduction by "gemmation," as the other is of reproduction by "fission." In this case the growing sarcode pushes outside the initial chamber until at a certain stage it builds a new wall around itself, while still maintaining connection with the parent cell. This second segment may give origin to a third, and so on until a colony is established, each offspring occupying an apartment added to the parental home. It is easy to see that the style of architecture of these tenements may be almost infinitely varied by varying the shape and position of these annexes. Each annex may have any of the forms of the monothalamous shells or any modification of them, and the arrangement may be in straight or curved lines, in concentric circles or planospiral coils, in single or double series spirally coiled, in two or three alternating series not spiral, or even in an irregular and disorderly mass.

Usually in the development of the polythalamous shell each successive segment uses the party walls of the preceding segments, so far as they may be available, in the construction of its own annex, but in some of the higher types of the hyaline series it will be found that each chamber has a complete wall of its own, thus making double partitions between them. In some of these higher types there is developed also what is called a "supplemental skeleton," which consists of a deposit of shell-substance on the outside of the original wall, thereby adding to its thickness, filling in the hollows between the segments and at the umbilici, and sometimes growing out into protuberances of various shapes. This supplemental skeleton is generally traversed by a set of canals or sinuses—passages left during the deposit of the shell-substance, and probably occupied by threads of sarcode during the life of the animal.

The separation into families, genera, and species of a group of animals like the foraminifera, where variation is the rule and passage from one type to another is by a sliding scale and not even by a series of steps, is extremely difficult, and must always remain unsatisfactory in some particulars; but for convenience of reference, if for no other reason, a classification of some sort is demanded, and various schemes, which it is unnecessary here to enumerate, have been put forth to bridge the difficulty. In all these schemes the primary divisions are founded upon the structure of the test as above described—that is, whether chitinous, arenaceous, or calcareous, and whether perforate or imperforate. Beyond these distinctions, which seem to have a physiological foundation, there is nothing upon which to base a classification but the form of the test, which, as we have seen, is never determinate enough to permit of the establishment of fixed boundary lines. Generic and specific names of foraminifera, therefore, must not be considered as having much zoological value, but only as convenient titles applied to certain typical forms around which many varieties may be grouped. And it must be remembered that, however elastic the definitions of species, or even genera, there will often be a margin of doubt, and the determination of place in the classification must be left to the preference of the individual observer.

A few words concerning the manipulation of material and specimens may be of assistance to those beginning the study of the foraminifera.

Collection of recent shallow-water forms may be made from shore sands, from the anchor and chains, and especially from the "chain lockers" of ships, from sponge sand, and by means of boat dredges from the shallow waters of the coast. Deep-water forms are only obtainable by special apparatus, such as is used in deep sea sounding or in purely scientific explorations of the ocean bed. The specimens may be freed from mud by the process of decantation—that is, repeatedly agitating in water, and, after a very brief period to allow subsidence of the shells, pouring off the turbid surface water. Or the material may be put in a with a good acromatic lens magnifying about 10 diameters, is most convenient. A small quantity of the material in a shallow watch glass blackened on the under side, being placed under the lens, is carefully inspected, and when a specimen is found which it is desired to preserve it may be readily removed by means of a very fine camel's-hair pencil slightly moistened between the lips. Transfer of specimens should be attempted with the moistened pencil only, as the use of forceps is certain to crush the delicate shells.

For preservation of the identified specimens in numbers for study nothing is better than wooden slides of regulation size—1 by 3 inches. These may have either a concavity drilled in one side nearly through the wood and painted black, or a hole bored entirely through the slide and one side covered with heavy blackened paper. A removable cover to this little cavity may be cut from a thin sheet of mica and held in place either by a spring clamp or by slipping it under the thin paper front of the slide, which is left unglued about the center for that purpose.

To make a section the specimen should be attached in the desired attitude to the face and near the end of a glass slip by means of the minutest drop of liquid glue. The attitude of the specimen must be carefully preserved until the glue has set. The shell is then covered with chloroform or xylol balsam, which may be made to penetrate the chambers of the shell and be rapidly hardened by the application of direct heat up to the boiling temperatnre. Superfluous balsam being cut away, the shell supported by the balsam is rubbed lightly upon a hone, kept thoroughly wet with water, until the desired section is exposed. The balsam is then dissolved away by chloroform, and the glue by water, and the specimen mounted.

The manner in which specimens shall be mounted will depend upon the preferences or ingenuity of the preparator, and the arrangements he may make for the storage of his collection. If a cover-glass is used it should not be sealed on, as the underside of the glass is almost certain to "sweat" sooner or later, and obscure the specimen. It may be worth while to say that for the attachment of the shells to any surface the author has not found anything better than microscopists' gold size. The best instrument for transferring the minute drop of adhesive material of whatever kind to the point where the shell is to be attached is the finest obtainable sewing needle, the eye end inserted in a slender handle and the point broken off at the thickest part of the needle.

The literature of the subject is very large, though most of it is to be found in journals of natural history and transactions of societies. With Carpenter's "Introduction to the Study of the Foraminifera," Brady's "Report on the Foraminifera collected by H. M. S. *Challenger*," and Sherborn's "Index to the Genera and Species of the Foraminifera," the student will be able to begin work in an intelligent manner and to find references to all that has been published on this subject up to the most recent date.

ANALYTICAL KEY TO FAMILIES.

Subkingdom PROTOZOA.—Body consisting of a minute mass of protoplasm, or an aggregation of such masses, without differentiation of parts into organs or tissues, either with or without a testaceous envelope or skeletal framework.

Class RHIZOPODA.—Protoplasmic body capable of protruding any portion of its substance in the shape of lobes, bands, or threads, for the purpose of locomotion or the prehension of food; generally more or less completely inclosed in a testaceous envelope; nucleus and contractile vesicle present or absent.

Order FORAMINIFERA.—Pseudopodia protruded as fine threads which flow together wherever they touch, forming a network of granular protoplasm; nucleus and vacuoles generally indistinguishable; tests either chitinous, calcareous, or of agglutinated sand or shells, never silicious.

Test chitinous, sometimes encrusted with foreign bodies.
  Aperture at one or both extremities .... Family I. GROMIDÆ.
Test arenaceous (composed of mud, sand, shells, or sponge spicules).
  Relatively large, one-chambered, or sometimes unsymmetrically segmented by constriction or adhesion, never truly septate...... Family II. ASTRORHIZIDÆ.
  Relatively small, usually regular in contour, one or many chambered; many-chambered forms sometimes imperfectly septate, often labyrinthic:
    Family III. LITUOLIDÆ.
Test arenaceous or calcareous.
  Segments in two or more alternating series, or spiral or confused, often dimorphous .... Family IV. TEXTULARIDÆ.
Test calcareous.
  Imperforate, porcellanous................................ Family V. MILIOLIDÆ.
  Perforate, hyaline.
    Chambers one, or many joined in a straight, curved, spiral, alternating, or branching series; aperture simple or radiate, terminal:
      Family VI. LAGENIDÆ.
    Chambers more or less embracing, following each other from the same end, or alternately at either end, or in cycles of three:
      Family VII. CHILOSTOMELLIDÆ.
    Chambers comparatively few, inflated, spirally arranged; apertures single or multiple, conspicuous .... Family VIII. GLOBIGERINIDÆ.
    Chambers typically spiral and rotaliform—all the segments visible on the upper side, those of the last convolution only on the lower (apertural) side. Aberrant forms evolute, outspread, acervuline, or irregular:
      Family IX. ROTALIDÆ.
    Chambers spiral or concentric; shell symmetrical, usually lenticular or discoidal .... Family X. NUMMULINIDÆ.

ANALYTICAL KEY TO GENERA.

Family I. GROMIDÆ.

Aperture single.
  Test large, ovate.
    Mouth central, in a depression at the broad end; test closely adherent to the body of the animal

Family II. ASTRORHIZIDÆ.

Walls thick, composed of sand or mud, slightly cemented..Subfamily ASTRORHIZINÆ.
Fusiform, branching, or flattened with angular or radiate margin; aperture at the end of each ray or branch ....Genus *Astrorhiza*.
More or less flask shaped or subcylindrical; aperture single, terminal: Genus *Pelosina*.
Subglobular, very irregular externally; apertures numerous, in horn-like protuberances..Genus *Storthosphæra*.
Columnar, branching, or irregularly outspread; adherent; apertures terminal: Genus *Dendrophrya*.
A rounded mass of radiating, branching tubes arranged in more or less distinct layers..Genus *Syringammina*.

Walls thick, composed of felted sponge spicules and fine sand, uncemented: Subfamily PILULININÆ.
Spherical; aperture a long, curved slit ....Genus *Pilulina*.
Subspherical, labyrinthic or cavernous, or having a central undivided cavity with subcavernous walls; no general aperture ....Genus *Crithionina*.
Oval or subcylindrical; aperture typically a rounded orifice at one end: Genus *Technitella*.
Cylindrical, long, slightly tapering, open at both ends.......Genus *Bathysiphon*.

Walls thin, composed of sand grains firmly cemented; test nearly spherical: Subfamily SACCAMMININÆ.
A single globular chamber, without general aperture.....Genus *Psammosphæra*.
A number of adherent globular chambers, without general aperture: Genus *Sorosphæra*.
One or several globular, pyriform or fusiform chambers, with or without tubular connection; apertures distinct ....Genus *Saccammina*.

Walls composed of firmly cemented sand grains, often mixed with sponge spicules; test tubular, sometimes imperfectly segmented..Subfamily RHABDAMMININÆ.
Elongate, tapering, simple; aperture at the broad end ....Genus *Jaculella*.
Elongate, cylindrical, simple or branched; aperture at one end, the other end rounded, sometimes inflated..................................Genus *Hyperammina*.
Fusiform or cylindrical, largely composed of sponge spicules; aperture at each end...Genus *Marsipella*.
Rectilinear, radiate or branching, with or without a central chamber; apertures at the open ends of the tubes ....Genus *Rhabdammina*.
Very variable, usually consisting of irregular inflated sacs, single or united; apertures multiple, tubulated..........................Genus *Aschemonella*.
Tubular, slender, flexible, simple or branched, chitino-arenaceous, in nonadherent masses ....Genus *Rhizammina*.
Tubular, branching, reticulated, adherent to the surface of shells or stones; apertures terminal ....Genus *Sagenella*.
Subcylindrical, adherent at one end, rounded at the other, constructed of loose sand grains; imperfectly septate ....Genus *Botellina*.
Columnar, straight or crooked, adherent by an expanded base, enlarging or branching toward the apex; aperture terminal ....Genus *Haliphysema*.

Family III. LITUOLIDÆ.

Test composed of coarse sand grains, rough externally......Subfamily LITUOLINÆ.
Not labyrinthic.
Test free.
Chambers one, or several united in a straight, curved, or irregular line, never spiral.......................................Genus *Reophax*.
Chambers numerous, partly or entirely spiral...Genus *Haplophragmium*.

Test composed of coarse sand grains, etc.—Continued.
Not labyrinthic—Continued.
Test adherent.
Chambers numerous, planoconvex....................Genus *Placopsilina*.
Labyrinthic.
Test free.
Chambers uniserial, straight or curved, never spiral..Genus *Haplostiche*.
Chambers partly or entirely spiral......................Genus *Lituola*.
Test adherent.
Chambers linear, vermiform, closely approximated; apertures a row of pores on each septal face ....Genus *Bdelloidina*.
Test composed of fine sand, smooth externally.........Subfamily TROCHAMMININÆ.
Chambers one.
Globular with several mammilate apertures ....Genus *Thurammina*.
Elongate, conical, with a large curved or irregular aperture at the basal extremity ....Genus *Hippocrepina*.
A single tube coiled upon itself in various ways; sometimes constricted, never truly septate ....Genus *Ammodiscus*.
Adherent, hemispherical, with or without a long slender tubular neck: Genus *Webbina*.
Chambers several.
United in a straight or curved line; rarely a single chamber: Genus *Hormosina*.
Rotaliform, nautiloid, or trochoid; more or less distinctly septate: Genus *Trochammina*.
Rotaliform; test composed of fusiform calcareous spicules..Genus *Carterina*.
Test relatively large, composed of fine sand; chambers arranged spirally or in concentric layers; walls cancellated......................Subfamily LOFTUSINÆ.
Lenticular or subglobular; chambers numerous, spiral, nautiloid: Genus *Cyclammina*.
Fusiform or subglobular, elongated axially; chambers spiral....Genus *Loftusia*.
Spheroidal, compressed; chambers in concentric layers.........Genus *Parkeria*.
Test more or less calcareous; distinctly septate; exclusively fossil: Subfamily ENDOTHYRINÆ.
Nodosariform; chambers sometimes slightly labyrinthic; aperture simple: Genus *Nodosinella*.
Cylindrical, attached by one end; chambers labyrinthic; aperture terminal cribrate...Genus *Polyphragma*.
Lenticular, consisting of a planospiral tube with a deposit of shell substance on both sides..Genus *Involutina*.
Nautiloid or rotaliform; aperture simple, at the inner margin of the final chamber..Genus *Endothyra*.
Nautiloid; aperture a number of pores on the face of the terminal chamber: Genus *Bradyina*.
Adherent; consisting of numerous subdivided segments, or of a mass of chamberlets...Genus *Stacheia*.

Family IV. TEXTULARIDÆ.

Test typically bi- or tri-serial; often dimorphous.........Subfamily TEXTULARINÆ.

Test typically bi- or tri-serial, etc.—Continued.

Monomorphous—Continued.

Segments alternating in three rows.

Aperture as in *Textularia* .... Genus *Verneuilina*.

Aperture simple, produced, central .... Genus *Tritaxia*.

Aperture porous .... Genus *Chrysalidina*.

Segments arranged spirally, with three chambers in each convolution.

Aperture partially covered by a valvular lip .... Genus *Valvulina*.

Dimorphous.

Early chambers biserial, later ones uniserial and rectilinear: Genus *Bigenerina*.

Early chambers small and biserial, later ones broadly arched and uniserial: Genus *Paronina*.

Early chambers planospiral, later ones biserial .... Genus *Spiroplecta*.

Early chambers triserial, later ones uniserial and rectilinear: Genus *Clavulina*.

Early chambers triserial, later ones biserial .... Genus *Gaudryina*.

Test typically spiral; sometimes bi- or tri-serial; aperture oblique, comma-shaped or some modification of that form .... Subfamily BULIMININÆ.

Monomorphous.

Spiral, elongate, more or less tapering, often triserial .... Genus *Bulimina*.

Much elongated, with a tendency to become asymmetrically biserial: Genus *Virgulina*.

Distinctly biserial, Textularian .... Genus *Bolivina*.

Biserial; aperture an arched or semicircular orifice with a vertical notch on the septal face of the last segment .... Genus *Pleurostomella*.

Dimorphous.

Early segments bulimine or virguline, later ones uniserial .. Genus *Bifarina*.

Test consisting of a double series of alternating segments, more or less coiled upon itself .... Subfamily CASSIDULININÆ.

Folded on its long axis, and coiled more or less completely upon itself: Genus *Cassidulina*.

Broad, arched on the dorsal side, slightly coiled .... Genus *Ehrenbergia*.

Family V. MILIOLIDÆ.

Test irregular, asymmetrical: aperture variable .... Subfamily NUBECULARINÆ.

Chamber one, inflated, adherent; aperture on the convex surface: Genus *Squamulina*.

Chambers more than one, in linear or very irregularly spiral series: Genus *Nubecularia*.

Test coiled on an elongated axis, in a single plane or inequilaterally; chambers two in each convolution .... Subfamily MILIOLININÆ.

Chambers in a single plane, embracing, the last two only visible: Genus *Biloculina*.

Chambers biloculine but subdivided in the interior .... Genus *Fabularia*.

Chambers in a single plane, all visible on both sides of the shell: Genus *Spiroloculina*.

Chambers inequilateral, coiled round the long axis of the shell so that more than two (usually three or five) are visible .... Genus *Miliolina*.

Test dimorphous; partly milioline, partly spiral or rectilinear: Subfamily HAUERININÆ.

Early chambers milioline, subsequently in a straight series .... Genus *Articulina*.

Early chambers partly milioline and partly planospiral, subsequently in a straight series .... Genus *Vertebralina*.

Early chamber an undivided planospiral tube, subsequently with two or more segments in each convolution .... Genus *Ophthalmidium*.

Test dimorphous, etc.—Continued.

Early chambers milioline, subsequently planospiral with more than two segments in each convolution........Genus *Hauerina*.

Chambers equitant, arranged as in Hauerina, the last convolution covering the previous whorls........Genus *Planispirina*.

Test planospiral or cyclical, sometimes crozier-shaped, bilaterally symmetrical: Subfamily PENEROPLIDINÆ.

Chamber one, an undivided planospiral tube........Genus *Cornuspira*.

Chambers numerous, undivided, planospiral or spiral at first and rectilinear or cyclical afterwards........Genus *Peneroplis*.

Chambers subdivided transversely; early segments embracing; arrangement wholly planospiral or partly cyclical........Genus *Orbiculina*.

Chambers subdivided into chamberlets; test discoidal........Genus *Orbitolites*.

Test spiral, elongated in the line of the axis of convolution..Subfamily ALVEOLININÆ.

Subglobular, elliptical, or fusiform........Genus *Alveolina*.

Test spherical; chambers in concentric layers........Subfamily KERAMOSPHÆRINÆ.

Chambers very numerous, irregularly shaped........Genus *Keramosphæra*.

Family VI. LAGENIDÆ.

Test monothalamous........Subfamily LAGENIDÆ.

A single undivided chamber........Genus *Lagena*.

Test polythalamous, straight, arcuate or planospiral........Subfamily NODOSARINÆ.

Monomorphous.

Straight or curved, circular in transverse section; aperture central: Genus *Nodosaria*.

Straight, compressed; aperture typically a narrow fissure..Genus *Lingulina*.

Compressed or complanate; segments V-shaped, equitant: Genus *Frondicularia*.

Straight or slightly curved, triangular or quadrangular in section: Genus *Rhabdogonium*.

Elongate, curved, circular in section; aperture marginal: Genus *Marginulina*.

Elongate, compressed or complanate; septation oblique; aperture marginal: Genus *Vaginulina*.

Vaginuline; septation very oblique; aperture a long slit down the ventral face of the final segment........Genus *Rimulina*.

Planospiral in part or entirely; complanate, lenticular, crozier-shaped or ensiform........Genus *Cristellaria*.

Dimorphous.

Early segments *Cristellarian*, later ones Nodosarian........Genus *Amphycoryne*.

Early chambers *Cristellarian*, later ones Linguline........Genus *Lingulinopsis*.

Early chambers *Cristellarian*, later ones Frondicularian....Genus *Flabellina*.

Early chambers *Frondicularian*, later ones Nodosarian: Genus *Amphimorphina*.

Early chambers *Rhabdogonian*, later ones Nodosarian....Genus *Dentalinopsis*.

Test polythalamous; segments arranged spirally around the long axis; rarely biserial and alternate........Subfamily POLYMORPHININÆ.

Monomorphous.

Segments bi- or tri-serial or irregularly spiral; aperture radiate: Genus *Polymorphina*.

Test irregularly branching................................Subfamily RAMULININÆ.
Composed of spherical or pyriform chambers, connected by long stoloniferous tubes..Genus *Ramulina*.

Family VII. CHILOSTOMELLIDÆ.

Segments oval, each springing from the base of the previous one and entirely enveloping it..Genus *Ellipsoidina*.
Segments oval, put on alternately at either end of the test.....Genus *Chilostomella*.
Segments alternating at three sides so as to leave exposed portions of two segments and the whole of the final one..........................Genus *Allomorphina*.

Family VIII. GLOBIGERINIDÆ.

Test a single spherical chamber perforated with large and small foramina:
Genus *Orbulina*.
Test rotaliform, trochoid or planospiral; segments few, inflated, coarsely perforated:
Genus *Globigerina*.
Test regularly nautiloid and involute; walls thin, finely perforated, spinous:
Genus *Hastigerina*.
Test regularly or obliquely nautiloid and involute; walls thick, smooth, very finely perforated...Genus *Pullenia*.
Test nearly globular, composed of a few coiled segments........Genus *Sphæroidina*.
Test trochoid, segments inflated, finely perforated; aperture consisting of rows of pores along the septal depressions.........................Genus *Candeina*.

Family IX. ROTALIDÆ.

Test spiral, nonseptate..................................Subfamily SPIRILLININÆ.
A complanate, nonseptate tube, free or attached ....Genus *Spirillina*.
Test spiral, septate, rotaliform; rarely evolute, very rarely irregular or acervuline:
Subfamily ROTALINÆ.
Conical; consisting of an external spiral or annular layer of chambers, the interior of the cone being filled with hyaline substance or by a mass of compressed chambers ....Genus *Patellina*.
Trochoid or complanate, spiral at the apex, later segments often annular or irregular; apertures opening into a deep central vestibule, or sometimes consisting of sutural pores or bordered foramina.............Genus *Cymbalopora*.
Trochoid or planoconvex, rarely complanate; rather coarsely porous; aperture an arched slit at the umbilical margin of the last segment, often protected by an umbilical flap ....Genus *Discorbina*.
Complanate; early segments spiral, later ones cyclical; apertures peripheral:
Genus *Planorbulina*.
Upper side usually more convex than the lower; very finely porous; aperture a large slit at the umbilical end of the inferior sutural margin of the last segment ....Genus *Pulvinulina*.
Lower side usually the more convex; very finely porous; aperture a neatly arched slit near the middle of the inferior sutural margin of the last segment:
Genus *Rotalia*.
Lower side usually the more convex; coarsely porous; aperture near the outer end of the final suture, sometimes with a phialine neck...Genus *Truncatulina*.
Nearly alike on the two faces; coarsely porous...............Genus *Anomalina*.
Lenticular, periphery furnished with radiating spines.........Genus *Calcarina*.
Convex or monticulate, adherent; segments few, spreading radially or superimposed; aperture at the end of the final segment.........Genus *Carpenteria*.
Columnar, adherent by a slightly spreading base; segments numerous, spiral; aperture at the inner margin of the final segment............Genus *Rupertia*.

Test consisting of irregularly heaped chambers............Subfamily TINOPORINÆ
Lenticular or subsphæroidal, with radiating marginal spines and tuberculated surface; chambers arranged in tiers on each side of a central planospiral disk: Genus *Tinoporus*.
Sphæroidal or spreading, without spines; free or adherent, structure acervuline, radiating or laminated; chambers rounded or polyhedral, coarsely perforated; no general aperture.......................................Genus *Gypsina*.
Planoconvex, spreading, adherent; chambers acervuline; wall finely perforated; apertures numerous, marginal................................Genus *Aphrosina*.
Columnar, branching, attached by the base; segments numerous crowded around the long axis; coarsely perforated; no general aperture: Genus *Thalamopora*.
Encrusting or branching, parasitic; surface areolated; color pink or sometimes white...Genus *Polytrema*.

Family X. NUMMULINIDÆ.

Bilaterally symmetrical; chambers extending from end to end and arranged in convolutions perpendicular to the long axis of the shell..Subfamily FUSULININÆ.
Fusiform or subglobular; chambers entire......................Genus *Fusulina*.
Subglobular, elongated or subcylindrical; chambers subdivided by secondary septa...Genus *Schwagerina*.
Bilaterally symmetrical, nautiloid.....................Subfamily POLYSTOMELLINÆ.
Supplemental skeleton absent or rudimentary; no external septal pores or bridges; aperture a curved slit............................Genus *Nonionina*.
Supplementary skeleton, septal bridges and canal system present: aperture a V-shaped line of perforations at the base of the septal face..Genus *Polystomella*.
Lenticular or complanate...............................Subfamily NUMMULITINÆ.
Lenticular, consisting of a coiled nonseptate tube embedded in a mass of shell substance..Genus *Archædiscus*.
Lenticular, spiral, inequilateral; chambers equitant, simple above, constricted into two portions below...............................Genus *Amphistegina*.
Complanate and planospiral, all the convolutions visible; chambers undivided: Genus *Operculina*.
Complanate and planospiral; chambers divided into chamberlets: Genus *Heterostegina*.
Lenticular, planospiral, equilateral; chambers equitant, each convolution nearly or quite enclosing all the previous ones....................Genus *Nummulites*.
Complanate, regular, equitant, but the alar prolongations thin and transparent, exposing the outlines of previous convolutions ....Genus *Assilina*.
Complanate with thickened center, or lenticular.......Subfamily CYCLOCLYPEINÆ.
Composed of a single layer of chambers arranged in concentric annuli, with superimposed laminæ of finely tubulated shell substance thickest at the center: Genus *Cycloclypeus*.
Composed of a single layer of concentric chambers, with superimposed layers of flattened chamberlets....................................Genus *Orbitoides*.

CATALOGUE.

Family II. ASTRORHIZIDÆ.

Subfamily ASTRORHIZINÆ.

Walls thick, composed of loose sand or mud, very slightly cemented.

Genus ASTRORHIZA.

Test fusiform or depressed. Depressed forms either sublenticular with angular or irregularly radiate margin, or in branching masses. Apertures at the end of each ray or branch.

ASTRORHIZA GRANULOSA Brady.

(Plate 1.)

Test fusiform, composed of fine gray sand rather loosely cemented; cavity a tube of nearly uniform diameter, open at both ends; extremities of the test often tinged brown. Section shows thickness of shell and dimensions of cavity. Length, 4.5 mm. ($\frac{3}{16}$ inch), more or less.

Locality.—North Atlantic (stations 2568, 2570, 2723), 1,685 to 1,781 fathoms.

ASTRORHIZA CRASSATINA Brady.

(Plate 2.)

Test elongate, irregularly cylindrical. Differs from *A. granulosa* in that the cavity is more or less constricted at uncertain intervals. Length, 6 mm. ($\frac{1}{4}$ inch) or more.

Localities.—North Atlantic off Georges Bank, off Long Island, and off Chesapeake Bay (stations 2570, 2586, 2723), 328 to 1,813 fathoms.

ASTRORHIZA ANGULOSA Brady.

(Plate 3, fig. 1.)

Test irregularly triangular, depressed, thick, fragile, composed of fine gray sand loosely coherent; cavity a central globular chamber with tubes radiating to the angles and terminating in simple apertures. Section to show the cavity.

Locality.—Marthas Vineyard (station 2569), 1,782 fathoms.

ASTRORHIZA ARENARIA Norman.

(Plate 3, fig. 2.)

Test compressed, radiate or branched, composed of fine gray sand loosely cemented; very fragile; cavity corresponds with the form of the test; aperture at the end of each ray or branch.

Localities.—Off Marthas Vineyard and Georges Bank (stations 2547, 2570, 2586), 328 to 1,813 fathoms.

Genus PELOSINA.

Test free, typically monothalamous; rounded, cylindrical, tapering or irregularly fusiform; walls composed of mud with a chitinous lining; aperture single, terminal.

PELOSINA VARIABILIS Brady.

(Plate 4, fig. 1.)

Specimens both cylindrical and flask-shaped, one of them consisting of two quite irregular chambers; walls composed of mud with an occasional adhering shell. Length, 3 to 6 mm. (1/8 to 1/4 inch). Much larger specimens are common.

Locality.—Gulf of Mexico (station 2395), 347 fathoms.

Genus STORTHOSPHÆRA.

Test subglobular, very irregular externally; interior smooth.

STORTHOSPHÆRA ALBIDA Schultze.

(Plate 4, fig. 2.)

Subglobular or ovoid; surface roughened by prominent, rather thin ridges and protuberances; wall of medium and variable thickness, composed of very fine sand loosely cemented; cavity rounded, smooth; no visible aperture; color very light gray. Diameter, about 1.5 mm. (1/16 inch).

Locality.—Gulf of Mexico (station 2385), 730 fathoms.

Subfamily PILULININÆ.

Test monothalamous; walls thick, composed chiefly of felted sponge spicules and fine sand, without calcareous or other cement.

Genus PILULINA.

Test nearly spherical; aperture a long and more or less curved slit.

PILULINA JEFFREYSII Carpenter.

(Plate 5.)

Test spherical, thin, fragile, composed of sponge spicules and fine sand; cavity undivided, smooth; aperture a narrow curved slit with slightly protuberent lips. Section shows the large smooth cavity with thin walls. Diameter varies from 1.25 to 3 mm. (1/20 to 1/8 inch).

Locality.—North Atlantic; station not recorded.

Genus CRITHIONINA.

Labyrinthic or cavernous, or having a central undivided cavity with subcavernous walls.

loosely aggregated; color grayish white; cavity smooth, with or without more or less numerous pits or depressions in the walls; no traces of septa; no visible aperture. Average diameter, about 1.5 mm. ($\frac{1}{16}$ inch).

Localities.—North Atlantic, off Marthas Vineyard and Block Island (stations 2584, 2586, 2221, 2234), 328 to 1,525 fathoms.

CRITHIONINA PISUM, variety HISPIDA, new.

(Plate 6, fig. 2.)

In form like *C. pisum*, but smaller; characterized by the bristly appearance of the surface, caused by the projection of great numbers of sponge spicules arranged for the most part nearly perpendicular to the surface of the test. The very hispid tests have thinner walls than those with fewer projecting spicules; texture of walls and shape of cavity same as *C. pisum*. No visible aperture.

Localities.—Southeast of Georges Bank, Gulf of Mexico, and coast of Oregon (stations 2570, 2571, 2379, 2394, 3080), 93 to 1,813 fathoms.

Genus BATHYSIPHON.

Test long, cylindrical, slightly tapering; in the form of a straight or curved tube open at both ends.

BATHYSIPHON RUFUM de Folin.

(Plate 7.)

Test long, very slender, tapering gradually, smooth and polished externally, rather conspicuously constricted at very irregular intervals along its whole length; color a rich reddish brown; walls of medium thickness, composed of fine sand firmly and evenly cemented; cavity corresponds to the external form, the constrictions being equally marked within and without; apertures simple and terminal. Length, 3 to 9 mm. ($\frac{1}{8}$ to $\frac{3}{8}$ inch); diameter, 0.375 mm. ($\frac{1}{65}$ inch) or less.

Localities.—Gulf of Mexico and off the coast of Brazil (stations 2385, 2760), 730 to 1,019 fathoms.

Subfamily SACCAMININÆ.

Chambers nearly spherical; walls thin, composed of firmly cemented sand grains or shells of foraminifera.

Genus PSAMMOSPHÆRA.

Test a single globular chamber without any general aperture, the pseudopodia issuing from interstitial orifices.

PSAMMOSPHÆRA FUSCA Shulze.

(Plate 8, fig. 1.)

Nearly spherical, free or adherent, rough, constructed of comparatively large white grains of sand firmly cemented in a single layer; cavity as smooth as the nature of the material will admit, but not lined with cement substance, nor are the angles between the sand grains smoothly filled; no general aperture; color of the cement substance light grayish brown. Diameter, about 1.5 mm. ($\frac{1}{16}$ inch).

Locality.—Off Havana (station 2343), 279 fathoms. A variety of this species, taken off the coast of South Carolina, has a test constructed of coarse black sand; the cement is light brown, as in the other.

PSAMMOSPHÆRA FUSCA, variety TESTACEA, new.

(Plate 8, fig. 2.)

Differs from the type principally in the composition of the walls, which are constructed of a single layer of dead shells of foraminifera. It is generally larger and very rough, resembling an accidental agglomeration of shells, but showing in section a smooth cavity, as in the strictly arenaceous forms.

Locality.—Found only in the Gulf of Mexico (stations 2358, 2383, 2399), 196 to 1,181 fathoms.

PSAMMOSPHÆRA PARVA (P. FUSCA Brady).

(Plate 9, fig. 1.)

Test free or adherent; spherical when free; when adherent having a smooth facet, usually with an incomplete wall on the attached side. Diameter, about 0.625 mm. ($\frac{1}{40}$ inch); walls thin, composed of fine sand firmly united, the cement substance filling in smoothly the interstices and angles of the sand grains, both externally and internally; test often built around a long sponge spicule, which transfixes the test, both ends of the spicule protruding; color deep reddish brown. This species is included with *P. fusca* by Brady, "Report on the Foraminifera," but the characters are quite distinct, and no intermediate forms have been found.

Locality.—Coast of Brazil (station 2760), 1,019 fathoms.

Genus SACCAMMINA.

SACCAMMINA SPHERICA M. Sars.

(Plate 9, fig. 2.)

Test globular or slightly pear shaped, smoothly and strongly built of medium-sized grains of sand; aperture a simple tubular opening in the more or less protuberant end of the shell. Diameter, about 1 mm. ($\frac{1}{25}$ inch).

Locality.—Off the Coast of Brazil (station 2760), 1,019 fathoms.

SACCAMMINA CONSOCIATA, new species.

(Plate 9, fig. 3.)

Free or adherent, subglobular; surface coarse and rough; walls thin, composed of rather coarse sand mixed with sponge spicules; color a rich reddish brown; orifices one or several, at the end of long slender tubes. Generally united into colonies, either in straight series, or curved, or confused, connected by stoloniferous tubes. Diameter of individual tests, 0.4 to 0.8 mm. ($\frac{1}{60}$ to $\frac{1}{30}$ inch).

Locality.—Off Bahia, Brazil (station 2760), 1,019 fathoms.

Subfamily RHABDAMMININÆ.

Test composed of firmly cemented sand grains, often with sponge spicules intermixed; tubular; straight, radiate, branched, or irregular; free or adherent, with one, two, or more apertures; rarely segmented.

Genus JACULELLA.

Test elongate, tapering; aperture at the broad end.

JACULELLA ACUTA Brady.

(Plate 9, fig. 4.)

Long, cylindrical, tapering, closed at the pointed end when perfect, open at the broad end; walls constructed of coarse sand; surface rough; color, light brown. Length, about 3 mm. ($\frac{1}{8}$ inch).

Locality.—Not recorded.

Genus HYPERAMMINA.

Test free or adherent; consisting of a long, simple or branching, arenaceous tube, the primordial end of which is closed and rounded; the opposite extremity, which is open and but little if at all constricted, forming the general aperture; interior smooth.

HYPERAMMINA FRIABILIS Brady.

(Plate 10, fig. 1.)

Test free, consisting of a long straight tube, one end closed and slightly inflated, the other end slightly contracted, forming a simple rounded aperture; cavity corresponds to the external form of the test;

walls thin, constructed of moderately fine sand, or sometimes almost entirely of sponge spicules.

Localities.—North Atlantic and the Gulf of Mexico (stations 2399, 2400, 2234, 2570), 200 to 1,800 fathoms.

HYPERAMMINA ELONGATA Brady.

(Plate 10, fig. 2.)

Long, straight, slender, cylindrical, the inferior extremity slightly inflated and closed, the oral end little if at all contracted; composed either of fine sand or of broken sponge spicules firmly cemented; color deep reddish brown. Differs from *H. friabilis* in the much smaller diameter of the cylinder, the relatively greater length, and the firmer walls.

Localities.—Gulf of Mexico, the North Atlantic 200 miles southeast of Marthas Vineyard, and the coast of Brazil (stations 2394, 2568, 2760, 2352, 2355, 2399), 190 to 1,781 fathoms.

HYPERAMMINA RAMOSA Brady.

(Plate 11, fig. 1.)

Test free, commencing as a globular, inflated chamber, continuing as a long, crooked, branching tube; walls composed of sand or of sand mixed with sponge spicules; color, light brown.

Localities.—Off Cape Hatteras and in the Gulf of Mexico (stations 2115, 2352, 2383), 463 to 1,181 fathoms.

HYPERAMMINA VAGANS Brady.

(Plate 11, fig. 2.)

Test commences in a spherical chamber and continues as a slender unbranched tube of nearly even diameter and of indefinite length; sometimes partly free, but for the most part wandering over the surface of fragments of shells of mollusks, or of foraminifera, in a confused, tortuous and aimless way, or coiled irregularly upon itself; walls thin, composed of fine sand; color brown.

Locality.—Gulf of Mexico (station 2399), 196 fathoms.

Genus MARSIPELLA.

Test fusiform or cylindrical, with an aperture at each end; largely composed of sponge-spicules, especially near the extremities.

MARSIPELLA ELONGATA Norman.

overlies the fundamental structure of sponge-spicules. Length, 3 to 4 mm. (1/8 to 1/6 inch).

Localities.—Caribbean Sea, Gulf of Mexico, and off Cape Fear (stations 2150, 2383, 2677), 382 to 1,181 fathoms.

Genus RHABDAMMINA.

Test rectilinear, radiate, or irregularly branching; with or without a central chamber; the open ends of the tubes forming the apertures.

RHABDAMMINA ABYSSORUM M. Sars.

(Plate 12, fig. 2.)

Test free, radiate, most commonly with three rays in the same plane, but occasionally with four or five or more rays sometimes projecting irregularly from the central body; walls thin; central chamber small; the tubular arms terminating in simple rounded apertures. The specimens exhibited are below the average in size, but were selected for convenience of mounting. Section shows the form of the cavity, and thickness of the walls.

Locality.—Gulf of Mexico (station 2383), 1,181 fathoms.

RHABDAMMINA DISCRETA Brady.

(Plate 13.)

Test in the form of a long, straight cylinder, slightly constricted at irregular intervals and open at both ends; cavity smooth; walls rather thin, constructed of coarse sand firmly cemented. Sometimes reaches a length of an inch or more.

Locality.—Off Chesapeake Bay (station 2731), 781 fathoms.

RHABDAMMINA LINEARIS Brady.

(Plate 14, fig. 1.)

Test free, long, straight or slightly bent, cylindrical, having an oval, inflated central chamber with two long arms projecting in opposite directions on the same line; tubular portion slightly tapering; walls vary in texture from very fine sand mixed with sponge-spicules to quite coarse angular sand-grains; cavity corresponds to the outward form of the test; apertures simple, one at each end. Length, 3 to 12 mm. (1/8 to 1/2 inch).

Localities.—Off Georges Bank, and off the coast of Brazil (stations 2570, 2760), 1,019 and 1,813 fathoms.

RHABDAMMINA CORNUTA Brady.

(Plate 14, fig. 2.)

Test free, asymmetrical, consisting of an inflated chamber of irregular contour, and numerous short arms radiating from the surface; walls thin, composed of a single layer of rather coarse grains of white sand, sometimes mixed with sponge spicules, firmly united by a brown cement substance; arms tubular, terminating in simple rounded apertures.

Localities.—North Atlantic, and the Caribbean Sea (stations 2115, 2150, 2234, 2571), 380 to 1,350 fathoms.

Genus RHIZAMMINA.

Unattached masses of fine, flexible, simple or branching chitino-arenaceous tubes.

RHIZAMMINA INDIVISA Brady.

(Plate 15, fig. 2.)

Slender, flexible, simple, chitinous tubes of a brownish color, thickly incrusted with small foraminifera (mostly Globigerina) and very fine sand. Test more or less contorted in drying; generally tapering toward the extremities; apertures terminal, simple. Length, 3 to 6 mm. ($\frac{1}{8}$ to $\frac{1}{4}$ inch).

Localities.—Southward of Long Island, the Straits of Yucatan, the Gulf of Mexico, and the coast of Brazil (stations 2234, 2355, 2380, 2760), 400 to 1,400 fathoms.

RHIZAMMINA ALGÆFORMIS Brady.

(Plate 15, fig. 1.)

Slender, chitinous tubes, incrusted with fine sand or small foraminifera; dichotomously branched; quite flexible while wet, very brittle when dry; found in tangled masses, from which it is extremely difficult to separate an unbroken specimen. Length, indefinite; may be an inch or more; diameter of tube, 0.12 to 0.3 mm. ($\frac{1}{200}$ to $\frac{1}{80}$ inch).

Locality.——ff the west coast of Mexico (station 3415), 1879 fathoms.

Family III. LITUOLIDÆ.

Test arenaceous, usually regular in contour and more or less definitely segmented; chambers frequently labyrinthic.

Subfamily LITUOLINÆ.

Test composed of coarse sand grains, rough externally; often labyrinthic.

Genus REOPHAX.

Test free; composed of a single flask-shaped chamber, or of several united in a straight, curved, or irregular line; never spiral.

mented; aperture a single, simple, round opening. Length, 0.35 to 75 mm. ($\frac{1}{70}$ to $\frac{1}{32}$ inch).

Localities.—Cape Hatteras, in the Gulf of Mexico, and off New York stations 2115, 2377, 2394, 2530, 2550, 2584), 400 to 1,000 fathoms.

REOPHAX DIFFLUGIFORMIS Brady, variety TESTACEA, new.

(Plate 16, fig. 1.)

Identical with the preceding, except that the test is much larger and omposed entirely of small empty shells of foraminifera. Section shows ie undivided chamber and the walls constructed of a single layer of hells.

Locality.—Southward of Long Island (station 2234), 810 fathoms.

REOPHAX SCORPIURUS Montfort.

(Plates 16, fig. 3; 17, fig. 1.)

Consists of a series of segments, few in number, irregular in shape, ined in a more or less curved or crooked line. The walls may be composed entirely of sand or of the shells of foraminifera, or in part of each.

Localities.—Off Marthas Vineyard, and southeast of Georges Bank stations 2221, 2570), 1,525 and 1,813 fathoms.

REOPHAX BILOCULARIS, new species.

(Plate 17, fig. 2.)

Composed of two segments united end to end in a straight or curved ine; primary segment oval, ovate, or cylindrical, constricted at the junction with the final segment, which is ovate, inflated, and terminates in tubular neck with a round orifice; walls composed of a single layer f shells of dead foraminifera, both small and large, mixed with fine and; surface often very irregular when large shells are built into the ralls. Length, about 1.5 mm. ($\frac{1}{16}$ inch).

This seems to be an intermediate form between *R. difflugiformis* and *R. scorpiurus.* Goës[1] figures a similar specimen under the name *R. nodulosus pygmæus*, but another specimen under the same name is igured having five segments. No example having more than two segments has been found among the hundreds taken from material dredged ff Cape Fear (station 2679), 782 fathoms.

REOPHAX PILULIFERA Brady.

(Plate 18, fig. 1.)

Segments three to five, inflated, rapidly increasing in size from the irst, forming a conical curved test; walls composed of coarse sand, ough; color, brown; aperture simple, terminal. Length, about 1.5 mm. $\frac{1}{16}$ inch).

Locality.—Off Bahia, Brazil (station 2760), 1,019 fathoms.

---

[1] Arct. and Scand. Foram.

REOPHAX DENTALINIFORMIS Brady.

(Plate 18, fig. 2.)

Test cylindrical, tapering, slightly curved, made up of four to six elongate, slightly inflated segments arranged in linear series. Walls composed of rather coarse sand, firmly cemented; aperture in the prolonged end of the terminal segment. Length, 1.5 to 3 mm. ($\frac{1}{16}$ to $\frac{1}{8}$ inch).

Locality.—Not recorded.

REOPHAX BACILLARIS Brady.

(Plate 18, fig. 3.)

Long, slender, cylindrical, straight or slightly bent, tapering gradually, composed of numerous segments (fifteen to twenty); sutures between the earliest segments indistinguishable, the later segments inflated and the sutures well marked; aperture simple, in the terminal segment; color, light gray. Length, 1.5 to 3 mm. ($\frac{1}{16}$ to $\frac{1}{8}$ inch).

Localities.—Nantucket Shoals, off Trinidad, south of Cuba, southeast of Marthas Vineyard, off Chesapeake Bay (stations 2041, 2221, 2228, 2568, 2723), 1,500 to 1,800 fathoms.

REOPHAX NODULOSA Brady.

(Plate 18, fig. 4.)

A long, cylindrical, tapering, straight or slightly bent test, composed of several (commonly six to ten) oblong or pyriform segments, arranged in linear series, slightly embracing; walls thin, arenaceous, smooth within and without; color, a rich brown; aperture simple, terminal. Section shows the smooth chambers and the thin embracing walls.

Locality.—Gulf of Mexico (stations 2385, 2395), 730 and 347 fathoms.

REOPHAX ADUNCA Brady.

(Plate 18, fig. 5.)

The distinguishing characteristics of this species are the inflated segments, their nearly equal diameter, and their irregular arrangement in a crooked line of succession. It is of smaller size and coarser structure than the other polythalamous species of Reophax.

Localities.—Off coast of Maryland and in Gulf of Mexico (stations 2228, 2338), 1,582 and 189 fathoms.

REOPHAX CYLINDRICA Brady.

(Plate 18, fig. 6.)

flat septal plates. Length, about 2.5 mm. ($\frac{1}{10}$ inch); diameter, 0.4 mm. ($\frac{1}{60}$ inch).

***Locality.*—A single specimen obtained about 200 miles southeast of Marthas Vineyard (station 2568), 1,781 fathoms.**

Genus HAPLOPHRAGMIUM.

Test free; partially or entirely spiral; nautiloid or crosier shaped; chambers numerous, not labyrinthic.

HAPLOPHRAGMIUM AGGLUTINANS d'Orbigny.

(Plate 19, fig. 2.)

Commences as a small, flat spiral of little more than a single convolution; continues as a straight series of cylindrical segments, gradually increasing in size; walls constructed of more or less coarse sand; surface rough, sutural lines indistinct; aperture central at the end of the final segment. Section shows form and arrangement of chambers.

***Localities.*—North Atlantic and Gulf of Mexico (stations 2041, 2115, 2385, 2374, 2568, 2576, 2679), 18 to 1,700 fathoms.**

HAPLOPHRAGMIUM CALCAREUM Brady.

(Plate 19, fig. 1.)

A large, coarse, compressed, falciform shell, with a short spiral portion and a more or less extended straight part, composed of two to six well-defined, broad segments; walls constructed of rather coarse coral sand neatly joined and firmly cemented; aperture simple, terminal. Length, about 3 mm. ($\frac{1}{8}$ inch).

***Locality.*—Arrowsmith Bank, Straits of Yucatan (station 2355), 399 fathoms.**

HAPLOPHRAGMIUM TENUIMARGO Brady.

(Plate 19, fig. 3.)

Test small, much compressed, the edges thin and jagged; segmentation obscure, early arrangement spiral, later rectilinear; walls of coarse sand; surface rough; aperture simple, terminal. Length, 0.75 to 1.5 mm. ($\frac{1}{32}$ to $\frac{1}{16}$ inch).

***Localities.*—Off Cape Hatteras and off Block Island (stations 2115, 2584), 843 and 541 fathoms.**

HAPLOPHRAGMIUM CASSIS Parker.

(Plate 19, fig. 4.)

Small, compressed, somewhat sigmoidal in outline, the edges rounded; segmentation obscure, early arrangement spiral, later arrangement linear, but the segments becoming broader and more and more diagonally placed; walls of coarse sand, but the surface comparatively

smooth; color light gray; aperture at the end of the final segment. Length, about 1.5 mm. ($\frac{1}{16}$ inch).

Locality.—Portland, Maine, 4 to 5 fathoms.

HAPLOPHRAGMIUM FOLIACEUM Brady.

(Plate 19, fig. 6.)

Flat on both sides and extremely thin, the early spiral convolutions quite distinct, the rectilinear segments broad and with sutural lines evident; walls smooth and built of rather coarse sand; color reddish-brown; aperture a terminal slit. Length, about 1.25 mm. ($\frac{1}{20}$ inch).

Localities.—Gulf of Mexico and off Marthas Vineyard (stations 2377, 2568), 210 and 1,781 fathoms.

HAPLOPHRAGMIUM EMACIATUM Brady.

(Plate 19, fig. 5.)

Thin, flat, nearly circular in outline, consisting of about two convolutions made up of numerous segments; lines of union of the segments more or less indistinct; walls composed of sand, or of sand and sponge spicules mixed, or sometimes almost wholly of broken sponge spicules arranged in an orderly manner parallel to the spiral axis of growth; color brown; aperture a transverse arched slit at the base of the final segment. Diameter, about 1 mm. ($\frac{1}{25}$ inch).

Localities.—West coast of Cuba, and off coast of Brazil, (stations 2352, 2760), 463 and 1,019 fathoms.

HAPLOPHRAGMIUM LATIDORSATUM Bornemann.

(Plate 20, fig. 1.)

A simple planospiral shell of about three convolutions, the segments rapidly increasing in size, the final convolution completely inclosing the others. Contour subglobular, septal lines distinct; aperture a slightly irregular transverse slit at the base of the final segment, with thin, well-formed lips; color grayish-brown. Diameter, about 1.5 mm. ($\frac{1}{16}$ inch). Section shows the arrangement of chambers, and the thick, rather coarsely arenaceous walls.

Localities.—Off Nantucket Shoals and Gulf of Mexico (stations 2041, 2352, 2385, 2586), 300 to 1,600 fathoms.

HAPLOPHRAGMIUM SCITULUM Brady.

(Plate 20, fig. 2.)

A planospiral shell of about three convolutions, somewhat flattened on both sides, depressed at the center, the outer convolution more or

Localities.—West coast of Cuba, south of Black Island, west coast of Patagonia, (stations 2352, 2584, 2784, 3080), 93 to 541 fathoms.

HAPLOPHRAGMIUM CANARIENSE d'Orbigny.

(Plate 20, fig. 3.)

Planospiral, much compressed, especially the earlier convolutions, the segments of the final convolution more or less inflated; structure coarsely arenaceous; surface rough; color reddish to grayish-brown; aperture a short transverse slit, with thin projecting lips, situated near the inner margin of the last segment. Diameter, about 1.25 mm. ($\frac{1}{20}$ inch).

Localities.—Off Nantucket shoals, south of Black Island, and coast of Oregon (stations 2251, 2584, 3080), 43 to 540 fathoms.

HAPLOPHRAGMIUM GLOBIGERINIFORME Parker and Jones.

(Plate 21, fig. 1.)

Has the same form as *Globigerina bulloides*, being composed of a series of gradually enlarging segments arranged spirally around a perpendicular axis, all the segments being visible on one face of the shell, and only the final convolution on the other. Walls composed of rather coarse sand, firmly and neatly cemented; color brown; aperture at the central margin of the final convolution. Size very variable.

Localities.—Off Nantucket Shoals, off Cape Hatteras, southeast of Marthas Vineyard, off coast of Brazil (stations 2041, 2115, 2568, 2760), 840 to 1,780 fathoms.

Genus HAPLOSTICHE.

Test free, uniserial, straight or arcuate; never spiral; chambers labyrinthic.

HAPLOSTICHE SOLDANII Jones and Parker.

(Plate 21, fig. 3.)

Elongate, cylindrical or tapering, rounded at the extremities, consisting of several (five to ten) chambers arranged in linear series; segments slightly embracing, lines of union indistinct; texture coarsely arenaceous; color light-gray; chambers subdivided by secondary septa; aperture porous or branched. Length, about 3 mm. ($\frac{1}{8}$ inch). Section shows the structure of the walls, the arrangement of the chambers and their labyrinthic character.

Localities.—Gulf of Mexico (stations 2377, 2399), 210 and 196 fathoms.

Subfamily TROCHAMMININÆ.

Test thin, composed of minute sand grains incorporated with calcareous or other inorganic cement, or embedded in a chitinous membrane; exterior smooth, often polished; interior smooth or (rarely) reticulated, never labyrinthic.

Genus THURAMMINA.

Test typically consisting of a single Orbulina-like chamber with several mammillate apertures.

THURAMMINA PAPILLATA Brady.

(Plate 22, fig. 1.)

Test spherical, with very thin walls constructed of fine sand grains firmly and smoothly cemented, inclosing a single undivided chamber. The surface is studded with more or less numerous nipple-like processes, each of which terminates in a simple aperture; color, various shades of brown. Diameter, 0.6 to 1.5 mm. ($\frac{1}{40}$ to $\frac{1}{16}$ inch).

***Localities.*—South of Long Island, Gulf of Mexico, southeast Georges Bank, coast of Brazil (stations 2225, 2383, 2385, 2570, 2760), 730 to 2,512 fathoms.**

THURAMMINA FAVOSA new species.

(Plate 21, fig. 2.)

Test spherical; walls very thin, arenaceous, brown; surface ornamented with a network of thin prominent ridges extending uniformly over the whole test, forming hexagonal pits; cavity smooth; apertures numerous, small, at the end of short tubular processes from some of the points of junction of the ridges. Diameter, about 0.8 mm. ($\frac{1}{30}$ inch).

***Locality.*—Gulf of Mexico (stations 2374, 2394), 26 and 420 fathoms.**

THURAMMINA CARIOSA new species.

(Plate 22, fig. 2.)

Spherical; surface rough, as if eroded; walls rather thick, cavernous; cavity globular, smooth; apertures not tubular; color a dirty brown. Differs from *T. favosa* in the thicker walls and coarser structure, the eroded rather than reticulated surface, the cavernous walls, and the nontubular orifices. Diameter, about 1 mm. ($\frac{1}{25}$ inch).

Locality.—Gulf of Mexico (stations 2385, 2394), 420 and 730 fathoms.

Genus AMMODISCUS.

Test free, formed of a tube coiled upon itself in various ways; sometimes constricted at intervals, never truly septate.

AMMODISCUS INCERTUS d'Orbigny.

Section shows a simple tube, without initial globular cavity, coiled upon itself in about twenty convolutions. Diameter, 0.75 to 3 mm. ($\frac{1}{32}$ to $\frac{1}{8}$ inch).

Localities.—Off coast of Maryland, south of Marthas Vineyard, Gulf of Mexico, coast of Brazil (stations 2171, 2243, 2383, 2385, 2586, 2760), 63 to 1,180 fathoms.

AMMODISCUS TENUIS Brady.

(Plate 23, fig. 1.)

A flattened disk, slightly, if at all, concave on the two faces, formed of a simple unconstricted tube of nearly uniform diameter coiled upon itself, each convolution slightly embracing the preceding. Differs from the last described species chiefly in the uniform size of the tube, and in the smaller number of convolutions. Diameter, about 2 mm. ($\frac{1}{12}$ inch).

Localities.—Off Cape Hatteras, off Nantucket Shoals, Gulf of Mexico, Panama Bay (stations 2115, 2352, 2385, 2395, 2805), 50 to 850 fathoms.

AMMODISCUS GORDIALIS Jones and Parker.

(Plate 24, fig. 1.)

Small, unsymmetrical in form, most often imperfectly lenticular; formed of a single tube of nearly uniform diameter coiled upon itself in varying directions. The degree of variation from the flat spiral differs with each specimen. Color light brown. Diameter, about 0.5 mm. ($\frac{1}{50}$ inch).

Localities.—Off Nantucket Shoals, southeast of Marthas Vineyard, and off coast of Oregon (stations 2041, 2568, 3080), 100 to 1,800 fathoms.

AMMODISCUS CHAROIDES Jones and Parker.

(Plate 24, fig. 2.)

Small, subglobular, formed of a narrow tube of uniform diameter coiled regularly in a series of superimposed layers, often terminating in a partial or complete convolution wound around the globular coil in a rectangular or diagonal direction; color brown; surface smooth and polished; aperture the open end of the tube. Diameter, 0.4 mm. ($\frac{1}{60}$ inch).

Localities.—Off Nantucket Shoals and coast of Oregon (stations 2041, 3080), 90 and 1,600 fathoms.

Genus WEBBINA.

Test adherent; consisting either of a single tent-like chamber, or of a number of such chambers connected by adherent stoloniferous tubes.

WEBBINA CLAVATA Jones and Parker.

(Plate 24, fig. 3.)

Test consists of either (1) the half of an oval or pear-shaped chamber, adherent to a bit of shell or other object which closes the flat side of the chamber, with a tubular prolongation of indefinite length also

adherent and incomplete; or of (2) a tube closed and inflated at one end, into the walls of which are built on all sides small foraminifera at rather close and irregular intervals. Texture finely arenaceous; color brown; aperture simple, terminal.

***Locality.*—Gulf of Mexico (stations 2352, 2385), 463 and 730 fathoms.**

Genus HORMOSINA.

Test consisting of a single rounded chamber, or, more usually, of several chambers in a single straight or arcuate series.

HORMOSINA GLOBULIFERA Brady.

(Plate 24, fig. 4.)

Consists of a single spherical chamber, or of several chambers (two to six), gradually increasing in size, and joined in a straight or slightly curved series; walls thin, of fine sand, neatly built, aperture simple at the end of a narrow tubular neck which terminates the final segment; color varies from white to reddish brown. Section shows the globular chambers, the thin walls, and the aperture leading to each successive chamber.

***Localities.*—Southeast of Georges Bank, and off coast of Brazil (stations 2530, 2570, 2760), 950 to 1,800 fathoms.**

HORMOSINA OVICULA Brady.

(Plate 25, fig. 2.)

Orbicular, oval, or pyriform segments, each having a more or less prolonged tubular neck, the segments arranged in a rectilinear series. Walls finely arenaceous, often rough externally with projecting sponge spicules incorporated with the sand. Length, 6 mm. ($\frac{1}{4}$ inch) or less.

Localities.—Gulf of Mexico and Atlantic off Cape Fear (stations 2383, 2399, 2677), 200 to 1,200 fathoms.

HORMOSINA CARPENTERI Brady.

(Plate 25, fig. 1.)

Pear-shaped segments, usually with a prolonged neck, nearly uniform in size, arranged in a curved or crooked series of indefinite length; walls finely arenaceous, firmly and smoothly cemented; aperture simple, terminal; color light brown. Section shows the thickness and structure of the walls, and the form of the chambers.

Locality.—Gulf of Mexico (stations 2382, 2383, 2385, 2398, 2400), 169 to 1,255 fathoms.

TROCHAMMINA PROTEUS Karrer.

(Plate 25, fig. 3.)

Test formed of a continuous tube, increasing slightly in diameter from the beginning, constricted at frequent and irregular intervals, coiled into the form of a disk, the convolutions being nearly in the same plane, or sometimes contorted into irregular forms. Diameter, 1.25 mm. ($\frac{1}{20}$ inch).

Localities.—Off west coast of Cuba, coast of Yucatan, Gulf of Mexico, Windward Islands, and coast of Brazil (stations 2352, 2355, 2394, 2750, 2760), 400 to 1,000 fathoms.

TROCHAMMINA LITUIFORMIS Brady.

(Plate 26, fig. 1.)

Consists of a simple tube, constricted at irregular intervals, coiled upon itself at the beginning either in planospiral convolutions or irregularly, subsequently becoming linear and more or less bent or contorted; surface smooth, color light brown; aperture terminal. Length, 5 mm. ($\frac{1}{5}$ inch) or less.

Localities.—Gulf of Mexico and coast of Brazil (stations 2352, 2394, 2395, 2760), 350 to 1,000 fathoms.

TROCHAMMINA CORONATA Brady.

(Plate 26, fig. 3.)

Test large, thick, biconcave, composed of numerous inflated segments arranged in a close spiral of three or more convolutions; walls distinctly arenaceous, even, but not smooth; sutures depressed; color pale brown or buff; aperture simple, terminal. Diameter, about 2 mm. ($\frac{1}{12}$ inch).

Locality.—Gulf of Mexico (station 2395), 347 fathoms.

TROCHAMMINA CONGLOBATA Brady.

(Plate 26, fig. 2.)

A tumid, subglobular shell, formed of a thin, irregularly segmented tube coiled upon itself in a constantly varying plane; segments much inflated, often transversely wrinkled; aperture the open, slightly constricted end of the tube; color brownish white. Diameter, about 1 mm. ($\frac{1}{25}$ inch).

Locality.—Gulf of Mexico (station 2395), 347 fathoms.

TROCHAMMINA RINGENS Brady.

(Plate 27, fig. 1.)

Test nautiloid, composed of a series of segments, rather rapidly increasing in size, arranged in planspiral convolutions, the final whorl completely inclosing the previous ones; contour ovoid, compressed,

adherent and in[illegible]
end, into the wall[illegible]
rather close and [illegible]
brown: aperture [illegible]
Locality.—G[illegible]

Test consisti[illegible]
several chambe[illegible]

Consists of [illegible]
[illegible]
[illegible]
[illegible]
[illegible]
chambers, [illegible]
chamber.
[illegible]

Family IV. TEXTULARIDÆ.

...ger species arenaceous, either with or without a per... basis; smaller forms hyaline and conspicuously per... bers arranged in two or more alternating series, or ...ed; often dimorphous.

Subfamily TEXTULARINÆ.

...or tri-serial; often bi-, rarely tri-morphous.

Genus TEXTULARIA.

...n two rows, alternating with each other; normal aperture ...t at the base of the inner wall of the final segment.

TEXTULARIA QUADRILATERA Schwager.

(Plate 28, fig. 3.)

..., compressed, tapering, quadrilateral, the two broader faces ...the angles prominent and sharp, both ends rounded; made ...double alternating series of segments to the number of seven, ...ss, in each row; aperture simple, near the base of the last ... structure hyaline and minutely perforate. Length, about ... inch).

...lity.—Specimens taken near Aspinwall, Isthmus of Panama (...n 2144), 896 fathoms.

TEXTULARIA TRANSVERSARIA Brady.

(Plate 28, fig. 4.)

...ongate, compressed, tapering, the broad faces convex, the angles ...; composed of a double row of chambers placed transversely to the ... axis of the shell, many of them open at the peripheral end, giving ...errated appearance to the edge of the test. Length, about 0.75 mm. (... inch).

Locality.—Off Carysfort Light, Florida (station 2641), 60 fathoms.

TEXTULARIA CONCAVA Karrer.

(Plate 28, fig. 5.)

Short, compressed, rapidly tapering, lateral faces flattened or concave, edges either square or rounded, angles full or rounded; texture rather roughly arenaceous; aperture a transverse arched slit with slightly protruding lips at the inner margin of the last segment. Length, about 1 mm. ($\frac{1}{25}$ inch). Readily distinguished from *T. quadrilatera* by the arenaceous texture of its walls.

Localities.—Off the island of Old Providence and off Carysfort, Florida (stations 2150, 2641), 382 and 60 fathoms.

TEXTULARIA CARINATA d'Orbigny.

(Plate 29, fig. 1.)

Short, triangular, compressed, the broad faces divided by a prominent ridge extending from the base toward the apex, the sutures strongly ribbed, the marginal angles acute, with irregular, short, rounded teeth, base quadrilateral, apex thin and slightly rounded; aperture a broad slit at the inner margin of the final segment; texture coarsely arenaceous. Length, 1.4 mm. ($\frac{1}{18}$ inch).

Locality.—Gulf of Mexico (station 2400), 169 fathoms.

TEXTULARIA RUGOSA Reuss.

(Plate 29, fig. 2.)

Pyramidal, with nearly equal sides, the angles rugged; segments rather thin, quadrangular, curved upon the flat, projecting at the sides and angles; sutural lines deep and arched. Length, about 1 mm. ($\frac{1}{25}$ inch).

Locality.—Specimen collected near the mouth of Exuma Sound, Bahamas (station 2629), 1,169 fathoms.

TEXTULARIA AGGLUTINANS d'Orbigny.

(Plate 29, fig. 4.)

Elongated, tapering, slightly flattened, composed of twenty segments, more or less, alternating in two rows, the later segments slightly inflated; texture rather coarsely arenaceous; aperture a smooth curved fissure on the inner side of the last segment. Length, about 1 mm. ($\frac{1}{25}$ inch).

Localities.—Near Aspinwall, Straits of Yucatan, Gulf of Mexico, coast of Brazil (stations 2144, 2358, 2385, 2760), 222 to 1,019 fathoms.

TEXTULARIA LUCULENTA Brady.

(Plate 29, fig. 3.)

Elongate, tapering, flattened; edges rounded; segments numerous; texture finely arenaceous. Length, about 2 mm. ($\frac{1}{12}$ inch).

Localities.—Near Old Providence, off Key West, Arrowsmith Bank (Yucatan) (stations 2150, 2315, 2355), 37 to 400 fathoms.

TEXTULARIA GRAMEN d'Orbigny.

(Plate 29, fig. 5.)

Subconical, compressed toward the tip, broadly oval at the base,

TEXTULARIA CONICA d'Orbigny.

(Plate 29, fig. 6.)

Small, short, conical, often a little compressed laterally, base quite flat; texture arenaceous; surface rough. Length, about 0.5 mm. ($\frac{1}{50}$ inch).

***Locality.*—Off Carysfort Light, Florida (station 2641), 60 fathoms.**

TEXTULARIA TROCHUS d'Orbigny.

(Plate 30, fig. 1.)

Short, conical, with a flat base and a rounded tip, in section circular both at the tip and base; walls thick and cavernous; texture rather coarsely arenaceous; aperture a narrow slit with smooth lips at the inner margin of the last segment. Length, about 1 mm. ($\frac{1}{25}$ inch).

***Localities.*—Off Cape Hatteras, west coast of Cuba, east coast of Florida (stations 2264, 2352, 2641), 60 to 460 fathoms.**

TEXTULARIA BARRETTII Jones and Parker.

(Plate 30, fig. 2.)

A large, symmetrical, elongated, conical shell, slightly compressed antero-posteriorly instead of laterally as in other species of this genus; texture arenaceous; surface smooth; sutures distinctly marked by narrow grooves; chamber cavities labyrinthic. Sections show the labyrinthic character of the chambers and the thick walls. Length, 4 mm. ($\frac{1}{6}$ inch), more or less.

***Locality.*—Off Little Bahama Bank (station 2655), 338 fathoms.**

Genus VERNEUILINA.

Test triserial, with textularian aperture.

VERNEUILINA PYGMÆA Egger.

(Plate 31, fig. 1.)

A short, conical test, composed of three series of segments arranged symmetrically around the long axis of the shell; segments inflated; walls finely arenaceous, smooth; aperture a long slit, with a slightly raised lower lip, at the inner margin of the final segment; color white. Length, about 1.5 mm. ($\frac{1}{16}$ inch).

***Locality.*—Gulf of Mexico (stations 2383, 2395), 1,181 and 347 fathoms.**

VERNEUILINA PROPINQUA Brady.

(Plate 31, fig. 2.)

Very similar in form to *V. pygmæa*, but is larger, coarser, rougher, less symmetrical, and in color a reddish brown. The aperture is without the raised lip seen in the other species.

Localities.—Specimens from four stations in the North Atlantic, one in South Atlantic, and two in the Gulf of Mexico (stations 2040, 2228, 2383, 2385, 2570, 2679, 2760), 730 to 2,226 fathoms.

Genus VALVULINA.

Test spiral, typically triserial, with three segments or rarely more in each convolution; free or adherent; aperture partially covered by a valvular lip.

VALVULINA CONICA Parker and Jones.

(Plate 31, fig. 3.)

Free or attached, short, conical; base broad and excavated; color brown, generally darker toward the apex; texture arenaceous; surface smooth; aperture at the inner margin of the terminal segment. One specimen shown is parasitic upon a fragment of *Rhabdammina;* the base of the attached *Valvulina* is surrounded by a border of fine white sand.

Localities.—Off Cape Hatteras, and south of Block Island (stations 2115, 2584), 843 and 541 fathoms.

Genus BIGENERINA.

Early chambers Textularian, later chambers uniserial and rectilinear.

BIGENERINA NODOSARIA d'Orbigny.

(Plate 31, fig. 4.)

The earlier segments, increasing rapidly in size, are arranged in two alternating series, forming the triangular flattened portion of the test, the remainder of the test is composed of three or four segments in a single straight series; aperture at the end of the final segment; texture coarsely arenaceous; surface rough. Length, about 1 mm. ($\frac{1}{25}$ inch).

Locality.—Off Carysfort Light, Florida (station 2641), 60 fathoms.

BIGENERINA ROBUSTA Brady.

(Plate 32, fig. 1.)

Test large, coarse, elongate, cylindrical, tapering slightly toward the initial end; textularian segments numerous, forming the greater part of the test; nodosarian segments few, sometimes irregular; aperture central in the final segment. Diameter, 2.5 mm. ($\frac{1}{10}$ inch) or more.

Locality.—Old Providence Island (station 2150), 382 fathoms.

BIGENERINA CAPREOLUS d'Orbigny.

BIGENERINA PENNATULA Batsch.

(Plate 32, fig. 2.)

Oblong, rounded at both ends, differing from *B. capreolus* only in the more arching form of the textularian segments, and the greater number (four or five) of segments in the linear series.

Locality.—Old Providence Island (station 2150), 382 fathoms.

Genus GAUDRYINA.

Early segments triserial (Verneuiline); aperture either textularian or situated in a short terminal neck.

GAUDRYINA PUPOIDES d'Orbigny.

(Plate 32, fig. 4.)

A small, subconical, symmetrical shell, about one-fifth of its length at the apex being formed of segments arranged triserially, the remaining portion composed of slightly inflated segments in double, alternating series; structure calcareous; surface smooth; aperture at the inner margin of the final segment. Length, about 0.825 mm. ($\frac{1}{30}$ inch).

Localities.—Off Nantucket Shoals, and southeast of Marthas Vineyard (stations 2041, 2568), 1,608 and 1,781 fathoms.

GAUDRYINA BACCATA Schwager.

(Plate 32, fig. 5.)

Differs from *G. pupoides* in that it is larger, less symmetrical, and the segments more inflated. It is especially characterized by the tendency to distortion produced by the occasional unsymmetrical outgrowth of one or more segments. Length, about 2 mm. ($\frac{1}{12}$ inch).

Localities.—Off Nantucket Shoals, south of Marthas Vineyard, off Block Island (stations 2040, 2221, 2570, 2584, 2586), 328 to 2,226 fathoms.

GAUDRYINA SUBROTUNDATA Schwager.

(Plate 33, fig. 1.)

Subcylindrical, tapering at the initial end; sutures depressed; aperture central, near the inner margin of the final segment; texture variable, the smaller specimens being comparatively fine and smooth, the larger coarse and rough. Length, 1 to 5 mm. ($\frac{1}{25}$ to $\frac{1}{5}$ inch).

Localities.—Specimens have been preserved from eight stations in the North and South Atlantic and the Gulf of Mexico (stations 2150, 2385, 2394, 2400, 2679, 2751, 2760, 2763), 169 to 1,019 fathoms.

GAUDRYINA FILIFORMIS Berthelin.

(Plate 33, fig. 2.)

Long, slender, tapering, smooth, the triserial portion very short, the biserial chambers numerous and symmetrically arranged; sutures well marked. Length, about 1.5 mm. ($\frac{1}{16}$ inch).

Locality.—Off west coast of Cuba (station 2352), 463 fathoms.

GAUDRYINA RUGOSA d'Orbigny.

(Plate 33, fig. 3.)

Elongate, triangular in section, the angles acute, triserial portion very short, biserial chambers alternately triangular and broadly quadrilateral in transverse section; structure coarsely arenaceous, compact. Length, 2 to 3 mm. ($\frac{1}{12}$ to $\frac{1}{8}$ inch). One specimen in the collection measures 4.5 mm. ($\frac{3}{16}$ inch).

Localities.—South of Marthas Vineyard and Gulf of Mexico (stations 2243, 2400), 63 and 169 fathoms.

GAUDRYINA SCABRA Brady.

(Plate 34, fig. 1.)

Resembles *G. pupoides* in form, but is larger, brown in color, coarsely arenaceous in texture, the sand sometimes mixed with sponge spicules; aperture a depressed slit at the inner margin of the last segment. Length, about 1.5 mm. ($\frac{1}{16}$ inch).

Localities.—Gulf of Mexico and west coast of Patagonia (stations 2352, 2385, 2784), 194 to 730 fathoms.

GAUDRYINA SIPHONELLA Reuss.

(Plate 34, fig. 2.)

Small, elongate, subcylindrical, occasionally distorted, the biserial segments numerous and somewhat inflated; aperture at the slightly projecting end of the final segment; color brown. Length, 0.5 to 0.8 mm. ($\frac{1}{50}$ to $\frac{1}{32}$ inch).

Locality.—Southeast of Marthas Vineyard (station 2568), 1,781 fathoms.

Genus CLAVULINA.

Early segments triserial, later ones uniserial and rectilinear; test generally either cylindrical or trifacial; aperture valvular.

CLAVULINA COMMUNIS d'Orbigny.

(Plate 34, fig. 3.)

Much elongated, cylindrical; the earliest portion triserial, conical, pointed, the remaining portion uniserial, straight, slightly depressed at the sutures; color, grayish-white; surface smooth or rough according as the walls are composed of fine or coarse calcareous sand; aperture round, at the end of a tubular projection from the final segment. Sections show the dimorphous character of the test, the thickness of walls,

CLAVULINA EOCÆNA Gümbel.

(Plate 35, fig. 1.)

Cylindrical or slightly tapering; triserial portion very short; nodosarian segments usually three or four in number, clearly defined by depressed sutures; walls coarsely arenaceous, rough; chambers partially divided by a network of incomplete septa springing from the outer wall; aperture a simple rounded orifice in a central slight depression at the end of the final segment. Section shows the apparent thickness of the walls due to the cancellar structure, and the form of the chambers. Length, about 1.5 mm. ($\frac{1}{16}$ inch).

Locality.—Gulf of Mexico (station 2377), 210 fathoms.

CLAVULINA PARISIENSIS d'Orbigny.

(Plate 35, fig. 2.)

The distinguishing characteristic of this species is the triangular contour of the triserial portion of the test; otherwise it strongly resembles *C. communis*. It is somewhat coarser and rougher than the latter, and near the oral end the sutures are often much depressed.

Locality.—Gulf of Mexico (stations 2315, 2377, 2385, 2400), 37 to 730 fathoms.

A variety collected near Key West, Florida, has a very rough test constructed of coral sand. (Plate 35, fig. 3.)

CLAVULINA PARISIENSIS, variety HUMILIS Brady.

(Plate 36, fig. 1.)

The variation consists in its smaller size, rougher exterior, the deep depression of the sutures, often forming a distinct neck between the two last segments, and the aperture borne at the end of a long tubular prolongation of the final segment.

Localities.—Gulf of Mexico, and off the coast of Brazil (stations 2377, 2399, 2400, 2762), 59 to 210 fathoms.

CLAVULINA ANGULARIS d'Orbigny.

(Plate 36, fig. 2.)

Arrangement of segments as in other species of *Clavulina*, triserial at first, then uniserial and rectilinear. Differs from the other species in the triangular contour of transverse section of the uniserial as well as the triserial portion of the test; aperture a central arched slit with a protruding lower lip.

Locality.—Straits of Yucatan (station 2358), 222 fathoms.

Subfamily BULIMININÆ.

Typically spiral; weaker forms more or less regularly biserial; aperture oblique, comma-shaped or some modification of that form.

...CULEATA d'Orbigny.

(Plate 37, fig. 4.)

...ntly compressed on three sides, segments ...ones bearing long slender spines, the ..., sometimes with short spines or slight

...all, Gulf of Mexico, southeast of Georges ...ions 2144, 2377, 2392, 2394, 2530, 2763), 210

...INA INFLATA Seguenza.

(Plate 37, fig. 5.)

...segments erect, short, and overlapping, the ...the segments crimped and sharply serrate.

...Mexico, southeast of Georges Bank, south of ...377, 2398, 2530, 2584), 210 to 956 fathoms.

Genus VIRGULINA.

...ited, with a tendency to become asymmetrically

...RGULINA SCHREIBERSIANA Czjzek.

(Plate 37, fig. 6.)

...cylindrical, slightly compressed on two sides, tapering ...tremities rounded, arrangement of segments irregularly ...g a twisted appearance to the shell; aperture a verti...al slit near the end of the last segment. Length, about ...inch.)

—Collected off Chesapeake Bay (station 2263), 430 fathoms.

VIRGULINA SUBSQUAMOSA Egger.

(Plate 37, fig. 7.)

...te-oval, compressed, margins rounded; segments overlapping, ...inflated, arranged in two inequilateral, alternating series; ...hin, tranparent, and finely perforated; aperture a loop-shaped ...the face of the last segment. Length, about 0.7 mm. ($\frac{1}{35}$ inch).

...lity.—Gulf of Mexico (station 2377), 210 fathoms.

Genus BOLIVINA.

...distinctly biserial, arrangement Textularian.

BOLIVINA ÆNARIENSIS Costa.

(Plate 37, fig. 8.)

Elongate, flattened, tapering, symmetrical; margins sharp and smooth; apex usually terminating in a spinous process; two or more delicate perpendicular ridges extending a variable distance from the apex toward the base; walls thin, transparent, minutely and profusely perforated; segments very regularly arranged in two alternating series; aperture loop-like at the inner margin of the last segment. Length, about 0.8 mm. ($\frac{1}{30}$ inch).

Localities.—Off Cape Hatteras, Gulf of Mexico, southeast of Georges Bank, south of Block Island (stations 2289, 2400, 2530, 2584), 7 to 956 fathoms.

BOLIVINA PUNCTATA d'Orbigny.

(Plate 38, fig. 1.)

Slender, elongate, tapering, rounded, symmetrical, slightly curved; composed of a double, alternating series of segments, twelve or more in each row; surface smooth and even; sutures not depressed; walls thin and finely perforated; aperture ovate, oblique, on the terminal face of the last segment. Length, about 0.8 mm. ($\frac{1}{32}$ inch).

Locality.—Not recorded.

BOLIVINA PORRECTA Brady.

(Plate 38, fig. 2.)

Straight, slightly tapering, nearly cylindrical in section; earlier segments in opposite alternating rows, later segments triangular and superposed, the sutures extending obliquely the whole breadth of the test; walls very thin, transparent and finely perforated; aperture large, oval, across the terminal face of the last segment. Length, about 1 mm. ($\frac{1}{25}$ inch).

Locality.—A single specimen from the North Atlantic, southeast of Georges Bank (station 2530), 956 fathoms.

Subfamily CASSIDULININÆ.

Test consisting of a Textularia-like series of alternating segments, more or less coiled upon itself.

Genus CASSIDULINA.

Test biserial, folded on its long axis, and coiled more or less completely on itself.

CASSIDULINA CRASSA d'Orbigny.

Locality.—Off head of Akutan Island, Alaska (station 2842), 72 fathoms.

CASSIDULINA SUBGLOBOSA Brady.

(Plate 38, fig. 4.)

Subglobular, the final segment slightly protruding, inequilateral, the segments being irregularly arranged; surface smooth; walls calcareous, imperfectly transparent, finely perforated; aperture an oval slit at the end of the last segment. Diameter, 0.8 mm. ($\frac{1}{30}$ inch).

Localities.—Gulf of Mexico, off Windward Islands, and Trinidad (stations 2383, 2751, 2754), 880 to 1,181 fathoms.

Family V. MILIOLIDÆ.

Test imperforate; normally calcareous and porcellanous, sometimes incrusted with sand.

Subfamily MILIOLININÆ.

Chambers two in each convolution, coiled on an elongated axis, either symmetrically in a single plane or inequilaterally. Aperture alternately at either end of the shell.

Genus BILOCULINA.

Chambers in a single plane, embracing; the last two only visible.

BILOCULINA BULLOIDES d'Orbigny.

(Plate 38, fig. 5.)

Oval, inflated, composed of a series of embracing segments applied alternately above and below the globular primordial chamber; walls thick, calcareous, soft; surface often incrusted with a thin layer of fine sand; aperture small, circular, on the more or less produced or tubular end of the last segment, usually bearing a small T-shaped valvular tooth. Length, about 1.25 mm. ($\frac{1}{20}$ inch). Transverse section shows the arrangement of the chambers.

Locality.—Off Havana, Cuba (station 2335), 204 fathoms.

BILOCULINA TUBULOSA Costa.

(Plate 39, fig. 1.)

In general characters like *B. bulloides*, except that the last two segments are separated by a deep groove on both sides. This groove may be so deep as to show the edge of the antepenultimate segment, and is often wider on one side than the other, so that the species passes by regular gradation into *Miliolina trigonula*. Length, 0.75 to 1.5 mm. ($\frac{1}{33}$ to $\frac{1}{16}$ inch).

Locality.—Specimens collected off the coast of Oregon (station 3080), 93 fathoms.

BILOCULINA RINGENS Lamarck.

(Plate 39, fig. 2.)

A stout, inflated, smooth, and polished shell, slightly compressed from above downward, nearly circular in outline when seen from above, the final segment projecting well beyond the preceding one, to which it is smoothly and firmly joined; aperture usually a broad slit with a nearly equally broad valvular lower lip. Diameter, 1.5 mm. ($\frac{1}{16}$ inch), more or less. Longitudinal section shows arrangement of chambers characteristic of the genus, and the apertures alternately at opposite ends of the shell.

Localities.—Off Cape Hatteras and in the Gulf of Mexico (stations 2115, 2352, 2385), 460 to 840 fathoms.

BILOCULINA COMATA Brady.

(Plate 39, fig. 3.)

Subglobular in form, otherwise like *B. ringens;* characterized specifically by surface ornamentation consisting of more or less conspicuous, fine, straight, parallel striæ covering the whole shell; aperture an arched slit, with a broad, thick valvular lower lip.

Locality.—West coast of Cuba (station 2352), 463 fathoms.

BILOCULINA ELONGATA d'Orbigny.

(Plate 39, fig. 4.)

Like *B. ringens* except that it is long oval in contour. The typical specimens are small, but there is constant variation both in size and breadth of oval.

Localities.—Gulf of Mexico and the North Atlantic (stations 2383, 2385, 2584), 500 to 1,200 fathoms.

BILOCULINA DEPRESSA d'Orbigny.

(Plate 40, fig. 1.)

Smooth, compressed, round; margin thin and sharp; aperture usually a long, narrow slit, with a valvular lower lip thinner and less prominent than in *B. ringens;* rarely the aperture is contracted to a nearly circular orifice. Longitudinal section shows the conformation and arrangement of the chambers.

Localities.—Gulf of Mexico and off Marthas Vineyard (stations 2374, 2378, 2568, 2570), 26 to 1,830 fathoms.

BILOCULINA DEPRESSA, var. SERRATA Brady.

Localities.—North Atlantic, Gulf of Mexico, and Panama Bay (stations 2530, 2383, 2399, 2805), 50 to 1,200 fathoms.

BILOCULINA DEHISCENS, new species.

(Plate 40, fig. 3.)

This species has the same general characters as *B. depressa* and its variety *serrata*, but the last two chambers are more or less separated at the sides, giving the shell the appearance of rupture from internal growth and distension. In general the separation is sufficient to show the sharp edge of the third segment on each side, but series presenting all degrees of gradation from Biloculina to Spiroloculina have been selected from material dredged at a single station.

Locality.—Gulf of Mexico (station 2377), 210 fathoms. (See *Spiroloculina robusta*, series.)

BILOCULINA LÆVIS Defrance.

(Plate 41, fig. 1.)

Less compressed than *B. depressa*, less inflated than *B. ringens*; characterized by the double border formed by the slight projection of the margin of the penultimate segment.

Locality.—Gulf of Mexico (station 2394), 420 fathoms.

BILOCULINA SPHÆRA d'Orbigny.

(Plate 41, fig. 2.)

Specific characters well marked. Contour nearly spherical; each chamber incloses the preceding one almost entirely, leaving exposed only a small circular segment of the penultimate chamber. Aperture an irregular, often branched or bordered, V-shaped slit. Section shows arrangement of chambers and degree of investment.

Localities.—North Atlantic, Gulf of Mexico, and coast of Brazil (stations 2352, 2385, 2415, 2754, 2760), 440 to 1,000 fathoms.

BILOCULINA IRREGULARIS d'Orbigny.

(Plate 41, fig. 3.)

Differs from other species of this genus in that it is compressed at the sides instead of from above downward. Seen from above the contour is oval; from the side the outline is broader, approaching the circular when the compression is considerable. Aperture circular or broad, with a valvular lip in the somewhat protuberant oral end.

Localities.—Caribbean Sea and Gulf of Mexico (stations 2754, 2144, 2352, 2355, 2383, 2385, 2394), 400 to 1,200 fathoms.

Genus SPIROLOCULINA.

Chambers arranged in a single plane, the whole of them visible on both sides of the shell.

SPIROLOCULINA ROBUSTA Brady.

(Plate 42, fig. 1.)

Much compressed laterally, broad oval to nearly round, more or less concave on both sides; extremities angular or pointed; periphery rounded, with sharp, projecting marginal angles, which are often toothed. Four to six segments visible on both sides, outlined by the acute prominent marginal angles; aperture round, with a T-shaped valvular tooth in the protruding end of the final segment. Longitudinal and transverse sections show the arrangement of chambers characteristic of the genus.

Locality.—Gulf of Mexico (stations 2383, 2399), 200 to 1,200 fathoms.

SPIROLOCULINA ROBUSTA, series.

(Plate 42, fig. 2.)

This is a series, selected from material dredged at a single station, to show an apparent evolution of *Spiroloculina robusta* from *Biloculina depressa*. The specimens are shown in pairs, the first of the pair being a whole shell resting upon its side, the other being a transverse section of a similar shell standing on end. The few specimens exhibited show a passage from one form to the other by well-defined steps, but with a large number of specimens the gradation is so easy that it becomes indefinable. The series illustrates the difficulties of classification in this order of animals.

Locality.—Gulf of Mexico (station 2377), 210 fathoms.

SPIROLOCULINA EXCAVATA d'Orbigny.

(Plate 41, fig. 5.)

Small, much compressed, long oval with projecting ends, very concave, showing the minute early segments; margins broad and rounded.

Locality.—Not recorded.

SPIROLOCULINA NITIDA d'Orbigny.

(Plate 41, fig. 4.)

More or less broadly oval, flat, thin, small; the segments inflated, without angles, the final one projecting at the oral end. Long diameter, about 0.75 mm. ($\frac{1}{32}$ inch).

Locality.—Collected in the Gulf of Tokyo.

SPIROLOCULINA LIMBATA d'Orbigny.

Localities.—Atlantic Coast of the Southern United States, and the Gulf of Mexico (stations 2312, 2313, 2358, 2420, 2614, 2641, 2400), 60 to 220 fathoms.

SPIROLOCULINA PLANULATA Lamarck.

(Plate 42, fig. 4.)

Compressed, broad oval to nearly circular; border rectangular or slightly rounded; segments not inflated, sutures rather indistinct; texture comparatively coarse. Diameter, about 1.25 mm. ($\frac{1}{20}$ inch).

SPIROLOCULINA ARENARIA Brady.

(Plate 43, fig. 1.)

Oval, much compressed, peripheral edge rounded, surface sandy and rough; sutural lines wholly obscured; aperture small, round, with a minute T-shaped tongue. Length, 0.75 mm. ($\frac{1}{32}$ inch).

Locality.—Off Carysfort Light, Florida (station 2641), 60 fathoms.

Genus MILIOLINA.

Chambers inequilateral, coiled around the long axis of the shell in such a way that more than two (usually three or five) are visible externally.

MILIOLINA SEMINULUM Linnæus.

(Plate 43, fig. 2.)

Contour as seen from above oval, from the end or side triangular, with rounded angles; surface smooth, with the clear white luster of porcelain characteristic of the Miliolidæ. Segments somewhat inflated, usually four of them partially visible on one side and three on the other; aperture round or oval, with a conspicuous appendicular tooth. Length, about 1.25 mm. ($\frac{1}{20}$ inch). The species is common in every latitude and at all depths.

Localities.—Specimens collected in the Gulf of Mexico, North Atlantic, and off coast of Brazil (stations 2570, 2568, 2754, 2383, 2392, 2760), 725 to 1,800 fathoms.

MILIOLINA GRACILIS d'Orbigny.

(Plate 43, fig. 5.)

Very small, long oval; segments nearly cylindrical, three of them visible on one face, and two on the other; aperture large with slightly protuberant lips. Length, about 0.5 mm. ($\frac{1}{50}$ inch).

Localities.—Cozumel Island, and off Carysfort Light, Florida (stations 2358, 2641), 222 and 60 fathoms.

MILIOLINA OBLONGA Montagu.

(Plate 43, fig. 3.)

Small, long oval in contour; otherwise like *M. seminulum.* Length, about 0.4 mm. ($\frac{1}{60}$ inch).

Localities.—From vicinity of the island of Trinidad and off coast of Brazil (stations 2754, 2760), 880 to 1,000 fathoms.

MILIOLINA AUBERIANA d'Orbigny.

(Plate 43, fig. 6.)

Larger than *M. seminulum*, and with the margins of the segments more sharply angular. Section shows the triangular contour of the shell and the characteristic milioline arrangement of the chambers.

Localities.—Atlantic coast of the United States, off the island of Trinidad, and off the coast of Brazil (stations 3150, 2570, 2584, 2754, 2760), 400 to 1,800 fathoms.

MILIOLINA CUVIERANA d'Orbigny.

(Plate 43, fig. 4.)

A rather large, smooth-shelled variety, characterized by the acutely angular margins of the five visible segments.

Localities.—Coast of Brazil (station 2762) and the Gulf of Tokyo, 59 and 9 fathoms.

MILIOLINA VENUSTA Karrer.

(Plate 44, fig. 2.)

Oval, angular, the margins of the three final segments extended so as to form well-marked keels; oval extremity of the last segment generally protuberant. Length, about 0.625 mm. ($\frac{1}{40}$ inch).

Localities.—West coast of Patagonia (station 2784) and Gulf of Tokyo, 194 and 9 fathoms.

MILIOLINA CIRCULARIS Bornemann.

(Plate 44, fig. 1.)

Smooth, slightly compressed, nearly circular in broadest outline; chambers inflated; aperture a crescentic slit with arched upper and thin projecting lower lip. Length, about 0.75 mm. ($\frac{1}{32}$ inch).

Localities.—Off Cape Hatteras, west coast of Cuba, and Trinidad (stations 2115, 2353, 2754), 167 to 880 fathoms.

MILIOLINA TRIGONULA Lamarck.

(Plate 44, fig. 3.)

Oblong, broad, oval in end view, exposing two chambers on one side and three, rarely four, on the other; oral end of the final segment often tubular; aperture round, with T-shaped valvular tooth. Transverse section of a broad specimen shows arrangement of the chambers.

Localities.—Atlantic coast of the United States and the Gulf of Mexico (stations 2228, 2570, 2385), 700 to 1,800 fathoms.

segment, the third by the free margin of the preceding segment. Aperture triangular, toothed. Length, about 0.625 mm. ($\frac{1}{40}$ inch).

Locality.—Off Windward Islands (station 2751), 687 fathoms.

MILIOLINA SUBROTUNDA Montagu.

(Plate 44, fig. 6.)

A small, thick, rounded, suborbicular shell, with three visible segments; surface slightly wrinkled transversely; terminal segment not projecting at the oral extremity; orifice large, with a prominent valvular tooth. Diameter, about 0.4 mm. ($\frac{1}{60}$ inch).

Locality.—Straits of Yucatan, 222 fathoms.

MILIOLINA VALVULARIS Reuss.

(Plate 44, fig. 5.)

A rather large, stout shell, having the same form and arrangement of segments as *M. circularis*. The distinguishing feature of this species is the aperture, which is a very narrow, irregularly bent, sometimes branching slit, with puckered lips.

Locality.—Cape Hatteras (station 2115), 843 fathoms.

MILIOLINA BUCCULENTA Brady.

(Plate 45, fig. 1.)

Large, subglobular, slightly and symmetrically compressed at the sides; especially characterized by the position of the three final and only visible segments in very nearly the same plane; aperture a long, arched slit across the face of the last segment.

Locality.—North Atlantic.

MILIOLINA LABIOSA d'Orbigny.

(Plate 45, fig. 3.)

Small, thin-shelled; segments few, inflated, often somewhat distorted, irregularly arranged; aperture large, crescent-shaped.

Locality.—Straits of Yucatan (station 2358), 222 fathoms.

MILIOLINA INSIGNIS Brady.

(Plate 45, fig. 2.)

Has the form of *M. trigonula*, or more often of *M. circularis*. The peculiarity of the shell is the surface ornamentation with fine, more or less prominent, parallel ribs.

Localities.—West coast of Cuba and the Caribbean Sea (stations 2352, 2150), 463 and 382 fathoms.

MILIOLINA UNDOSA Karrer.

(Plate 45, fig. 4.)

In this species of Miliolina the exposed portions of the segments are angular instead of being rounded or having a single sharp margin. Moreover, the angles of the segments are wavy, giving a crumpled appearance to the shell.

Locality.—Straits of Yucatan (station 2358), 222 fathoms.

MILIOLINA ANGULARIS, new species.

(Plate 46, fig. 1.)

An angular variety of Miliolina resembling *M. undosa*, except that the visible angles of the segments are very nearly right angles, slightly ribbed at the edges, and not sinuate.

Locality.—Straits of Yucatan (station 2358), 222 fathoms.

MILIOLINA BICORNIS Walker and Jacob.

(Plate 46, fig. 2.)

Oval, compressed, the final segment projecting posteriorly well beyond the preceding segment, and generally produced into a tubular neck anteriorly. The whole surface is striate, with rather fine, parallel, raised lines. Aperture round and toothed.

Localities.—Straits of Yucatan and coast of Florida (stations 2358, 2641), 60 to 222 fathoms.

MILIOLINA LINNÆANA d'Orbigny.

(Plate 46, fig. 3.)

Contour oval, compressed, much the same as *M. bicornis;* the surface marked with a few thick, irregular costæ in place of the striæ characteristic of the latter species.

Localities.—Straits of Yucatan, the Gulf of Mexico, coast of Florida (stations 2358, 2315, 2370, 2641, 2629), 13 to 222 fathoms.

MILIOLINA SEPARANS Brady.

(Plate 46, fig. 6.)

A single specimen of this species has been found. It has much the appearance of two small *Miliolina linnæana* grown together at the side,

MILIOLINA PULCHELLA d'Orbigny.

(Plate 46, fig. 4.)

Apparently a modified form of *M. linnæana*, in which the longitudinal costæ are somewhat less prominent, and are supplemented by quite numerous short diagonal ridges. Length, about 1 mm. ($\frac{1}{25}$ inch).

Locality.—Off Carysfort Light, Florida (station 2641), 60 fathoms.

MILIOLINA RETICULATA d'Orbigny.

(Plate 46, fig. 5.)

The single feature which characterizes this species is the surface ornamentation formed by two sets of fine, parallel striæ running diagonally to each other, producing a network of ridges.

Locality.—Straits of Yucatan (station 2358), 222 fathoms.

MILIOLINA AGGLUTINANS d'Orbigny.

(Plate 47, fig. 2.)

Broad, oval, thick, rounded; sutures obscure; terminal segment not produced; aperture large, with conspicuous appendicular tooth. Differs from *M. seminulum* in that the whole surface is incrusted with fine white sand. Length, from 1.5 to 0.6 mm. ($\frac{1}{16}$ to $\frac{1}{40}$ inch).

Locality.—Straits of Yucatan (station 2358), 222 fathoms.

Subfamily HAUERININÆ.

Test dimorphous; chambers partly miliolitie, partly spiral or rectilinear.

Genus ARTICULINA.

Chambers milioline at the commencement, subsequently in a straight series.

ARTICULINA SAGRA d'Orbigny.

(Plate 47, fig. 1.)

Irregularly long oval, or linear, compressed; the earlier segments milioline or confused, the later rectilinear; sutures constricted, each segment of the linear series overhanging the preceding; surface ornamented with fine, parallel, longitudinal striæ; aperture a long oval slit, with strongly everted lips, occupying the whole breadth of the oral extremity of the shell.

Localities.—Straits of Yucatan and the mouth of Exuma Sound (stations 2358, 2629), 222 to 1,169 fathoms.

Genus VERTEBRALINA.

Early chambers partly milioline and partly planospiral; later segments in straight series.

PENEROPLIS PERTUSUS Forskal.

(Plate 48, fig. 4.)

This species includes a wide variety of forms presenting all the intermediate stages from thick, slightly compressed, nautiloid shells, to the long, cylindrical, crosier-shaped varieties, and from these to the thin, compressed, rapidly widening forms. In all varieties the chambers are without divisions or constrictions, the apertures are porous, and the surface, with few exceptions, is striate.

Localities.—Straits of Yucatan and Exuma Sounds (stations 2352, 2629), 463 and 1,169 fathoms.

PENEROPLIS PERTUSUS, variety DISCOIDEUS, new.

(Plate 49, fig. 1.)

In this variety the final chambers completely surround the primary convolutions, forming a circular, thin disk resembling the discoidal forms of Orbulina, but distinguished by the entire absence of septa in the individual chambers.

Locality.—Key West Harbor; shallow water.

Genus ORBICULINA.

Chambers subdivided by transverse secondary septa; early segments embracing; arrangement either planospiral throughout or partly cyclical; contour nautiloid, auricular, crosier-shaped, or complanate.

ORBICULINA ADUNCA Fichtel and Moll.

(Plate 50, fig. 1.)

The only species of the genus. A planospiral, porcellanous, imperforate, polished shell, varying in contour from crosier-shaped to discoidal; surface usually pitted with minute depressions; the early convolutions embracing; chambers narrow and regularly subdivided; apertures a series of pores in two or more rows on the outer edge of the final chamber. It is distinguished from Peneroplis by the divided chambers, and from Orbitolites by the embracing early convolutions.

Localities.—Key West and St. Thomas; shallow water.

Genus ORBITOLITES

Test discoidal; either spiral (non-embracing) at the commencement, or with one or more inflated primordial chambers; subsequently cyclical; chambers more or less regularly divided into chamberlets.

length until they become concentric rings; segments divided by radial partitions into numerous chamberlets with free communication; a single row of pores on the margin of the disk forms the only exterior aperture.

Localities.—Key West, Florida, and off Cape Fear (station 2623).

ORBITOLITES DUPLEX Carpenter.

(Plate 51, figs. 2, 3.)

Shell thin, discoidal, slightly biconcave; primordial chamber conspicuous, globular; second chamber nearly surrounds the first; succeeding segments rapidly lengthen and quickly become annular. Chambers divided by septa into chamberlets, arranged in a double tier, with free communication. Peripheral orifices in two rows, corresponding to the double tier of chamberlets. This latter feature together with the early annular segments, distinguish this species from others of the genus. Diameter, 1 to 2.5 mm. ($\frac{1}{25}$ to $\frac{1}{10}$-inch).

Locality.—Key West, Florida.

ORBITOLITES TENUISSIMA Carpenter.

(Plate 52.)

An extremely thin and delicate shell, having the form of a circular disk with flat surfaces. In the arrangement of the chambers it commences as a convoluted, planospiral, nonseptate tube; it continues with a short series of spiral chambers and ends with a broad series of annular chambers. The spiral and annular chambers are partially divided, by partitions projecting from the inner walls, into numerous chamberlets. The chamberlets of each annulus communicate not only with each other but also with those of the succeeding annulus. A single row of pores opens on the margin of the final chamber. Diameter, from 1 to 20 mm. ($\frac{1}{25}$ to $\frac{4}{5}$ inch). The shaded portions in the figure are those parts of the specimen still occupied by the protoplasmic substance of the animal.

Locality.—Atlantic, south of Marthas Vineyard (station 2716), 1631 fathoms.

Family VI. LAGENIDÆ.

Test calcareous, very finely perforated; ether monothalamous, or consisting of a number of chambers joined in a straight, curved, spiral, alternating or (rarely) branching series. Aperture simple or radiate, terminal. No interseptal skeleton nor canal system.

Subfamily LAGENINÆ.

Test consists of a single chamber, either with or without an internal tube.

Genus LAGENA.

Test monothalamous, with either an external or internal tubular neck.

LAGENA GLOBOSA Montagu.

(Plate 53, fig. 4.)

Spherical, with a short conical protuberance ornamented with longitudinal costæ, body smooth, walls transparent, finely perforated, aperture leading into a short internal neck (entosolenian). This description applies to a single specimen from the Caribbean Sea near Aspinwall (station 2144), 896 fathoms.

LAGENA LONGISPINA Brady.

(Plate 53, fig. 2.)

Subglobular or pear-shaped; surface smooth; walls thin, glassy, more or less transparent, finely perforated, furnished with several (two to six or more) long, slender spines springing from the base of the shell; aperture round, central, at the apex, opening into a long neck or tube extending into the interior of the shell and terminating in a broadly expanded margin. Length of body, about 0.6 mm. ($\frac{1}{40}$ inch).

Localities.—Near Aspinwall, Gulf of Mexico, off Trinidad (stations 2144, 2394, 2754), 420 to 898 fathoms.

LAGENA GRACILLIMA Seguenza.

(Plate 53, fig. 3.)

A very delicate shell, with thin, transparent, and fragile walls and smooth surface; body either cylindrical or fusiform, drawn out at each end into a long thin neck; apertures simple, terminating the tubular neck at both ends of the shell, often surrounded at one end by an everted lip like the mouth of a phial.

Localities.—Various stations along the Atlantic and Gulf coast of the United States, at depths from 210 to 1,781 fathoms.

LAGENA ELONGATA Ehrenberg.

(Plate 53, fig. 1.)

Like *L. gracillima,* except that the body is long and cylindrical, with a short taper at both ends. Length, about 2 mm. ($\frac{1}{12}$ inch).

LAGENA DISTOMA Parker and Jones.

(Plate 53, fig. 5.)

Like *L. Gracillima* in its variety of forms, but characterized by more or less numerous, delicate, longitudinal striæ marking its surface.

LAGENA LÆVIS Montagu.

is opaque and with a roughened surface; aperture simple, at the end of the tubular neck. Diameter, about 0.6 mm. ($\frac{1}{40}$ inch).

Locality.—Not recorded.

LAGENA HISPIDA Reuss.

(Plate 53, fig. 8.)

Body globular or oval, with a long tubular neck projecting from one or both ends, the whole surface covered with fine, short, closely set spines. Length of body, about 0.4 mm. ($\frac{1}{64}$ inch).

Localities.—Gulf of Mexico, and off Windward Islands (stations 2398, 2751), 227 and 687 fathoms.

LAGENA SULCATA Walker and Jacob.

(Plate 53, fig. 7.)

Minute, flask-shaped; the neck long and slender, or short and stout, variously ornamented; the body decorated with numerous parallel, rather thin and sharp ridges or costæ. Length, about 0.4 mm ($\frac{1}{60}$ inch).

Localities.—Off Atlantic coast of the southern United States (stations 2420, 2614, 2641), 60 to 168 fathoms.

LAGENA STAPHYLLEARIA Schwager.

(Plate 54, fig. 1.)

Compressed pyriform, smooth, the apical margin rounded, the basal margin thin, broad and extended into four or five short stout spines; external aperture leading into an internal tube (entosolenian). Length, about 0.4 mm ($\frac{1}{60}$ inch).

Locality.—Caribbean Sea, near Aspinwall (station 2144), 896 fathoms.

LAGENA MARGINATA Walker and Boys.

(Plate 54, fig. 2.)

Contour round, lenticular, margin thin, sharp, and prolonged into a more or less broad wing projecting from the entire circumference; surface smooth; walls thin, generally transparent, and finely perforated; aperture a short horizontal slit at the margin, communicating with a tubular neck extending into the cavity of the shell. Diameter, about 1 mm. ($\frac{1}{25}$ inch).

Localities.—Caribbean Sea, Gulf of Mexico, and South Atlantic (stations 2144, 2150, 2385, 2394, 2395, 2754), 347 to 896 fathoms.

LAGENA CASTANEA, new species.

(Plate 54, fig. 3.)

Contour nearly circular, compressed, slightly protuberant at the oral end; margin rounded and smooth except at the aboral end, which is bicarinate; keels or wings thin, comparatively wide and well separated, extending about half the circumference of the test, joining each other at the extremities; mouth short, oval, with a contorted internal tubular neck. Diameter, about 0.5 mm. ($\frac{1}{50}$ inch).

Locality.—Near Aspinwall (station 2144), 896 fathoms.

LAGENA ORBIGNYANA Seguenza.

(Plate 54, fig. 4.)

Oval, compressed, the oral end protuberant and tapering; body smooth, the circumference bordered by three parallel wings or keels, the middle one widest. The aperture is at the end of a prolongation of the middle keel only. Diameter, about 0.5 mm. ($\frac{1}{50}$ inch).

Localities.—Caribbean Sea, Gulf of Mexico (stations 2117, 2144, 2355, 2394), 399 to 896 fathoms.

LAGENA CASTRENSIS Schwager.

(Plate 54, fig. 5.)

Form and general characters the same as *L. orbignyana;* distinguished by a surface ornamentation of regular rows of thickly set circular pits covering more or less completely the body and wings of the shell.

The published descriptions of *L. castrensis* call for a surface ornamentation of "exogenous beads," but in the specimens here described the surface is unquestionably pitted. The test is tricarinate and has all the other general characters of *L. castrensis.*

Locality.—Off Nantucket Shoals (station 2252), 38 fathoms.

Subfamily NODOSARINÆ.

Test polythalamous; straight, arcuate, or planospiral.

Genus NODOSARIA.

Test straight or curved, circular in transverse section; aperture typically central.

NODOSARIA ROTUNDATA Reuss.

(Plate 54, fig. 6.)

Oval or ovate, smooth, consisting of a few overlapp ng segments; sutures not depressed, indistinct; walls thin and white; aperture composed of a large number of radiating fissures, central at the end of the slightly produced terminal segment. Length, about 1 mm. ($\frac{1}{25}$ inch).

Localities recorded.—Five stations in the North Atlantic (stations 2212, 2550, 2571, 2577, 2586), 32 to 1,356 fathoms.

by the spines (one or several) projecting from the inferior end of the shell.

Localities.—Gulf of Mexico and west coast of Patagonia (stations 2352, 2377, 2395, 2784), 194 to 463 fathoms.

NODOSARIA RADICULA Linnæus.

(Plate 55, fig. 1.)

Oval, elongated, smooth, composed of two or more segments in a straight series; sutures depressed; aperture central, consisting of radiating fissures in the protuberant end of the last segment. Length, about 1 mm. ($\frac{1}{25}$ inch).

Localities.—South of Long Island, southeast of Georges Bank (stations 2234, 2570), 810 to 1,813 fathoms.

In typical specimens the segments are more inflated and the sutures more depressed than those figured in the accompanying plate.

NODOSARIA SIMPLEX Silvestri.

(Plate 55, fig. 2.)

Consists of two inflated, subglobular segments, the first terminating in a short spine, the second slightly elongated and tapering to the radiate aperture; sutures a little depressed; walls thin and transparent, finely perforated. Length, about 0.8 mm. ($\frac{1}{32}$ inch).

Locality.—Off Cape Hatteras (station 2115), 843 fathoms.

NODOSARIA PYRULA d'Orbigny.

(Plate 55, fig. 4.)

A long, slender, delicate shell, composed of a series of oval or ovate segments of nearly uniform size, joined together in a straight or slightly curved line by means of long tubular necks; surface smooth, without ornamentation. Length, indefinite. Owing to the fragility of the shell a whole one is rarely found. One specimen in the collection is over 8 mm. ($\frac{5}{16}$ inch) long.

Locality.—Gulf of Mexico (stations 2377, 2378, 2399), 68 to 210 fathoms.

NODOSARIA FARCIMEN Soldani.

(Plate 55, fig. 5.)

An elongated, tapering, slightly curved shell, composed of from four to eight oval or inflated segments, rapidly increasing in size from the first; segments separated by deep depressions, sometimes lengthened into a short neck; surface generally smooth, occasionally roughened about the sutures. Length, about 2.5 mm. ($\frac{1}{10}$ inch).

Localities.—Caribbean Sea, Gulf of Mexico, east coast of Florida (stations 2150, 2352, 2377, 2679), 210 to 782 fathoms.

NODOSARIA FILIFORMIS d'Orbigny.

(Plate 55, fig. 6.)

Long, slender, slightly curved, composed of numerous oval, smooth segments joined in linear series; sutures moderately depressed and transverse. Length, 3 to 4.5 mm. ($\frac{1}{8}$ to $\frac{3}{16}$ inch).

Locality.—Gulf of Mexico (stations 2377, 2378, 2399, 2400), 68 to 210 fathoms.

NODOSARIA CONSOBRINA, variety EMACIATA Reuss.

(Plate 56, fig. 1.)

Long, slender, slightly curved and tapering, composed of numerous short, nearly cylindrical segments, arranged in linear series; sutures not depressed except near the oral end; surface smooth. Length, 3 to 8 mm. ($\frac{1}{8}$ to $\frac{1}{3}$ inch).

Locality.—Gulf of Mexico (stations 2378, 2399), 68 and 196 fathoms.

NODOSARIA SOLUTA Reuss.

(Plate 56, fig. 3.)

A rather stout shell, composed of globular or short-oval segments, comparatively few in number, arranged in a straight or slightly curved line; initial segment large and spherical; surface smooth, or sometimes bristly rough about the sutures; aperture a round opening with short radiating fissures in the center of the protruding end of the terminal segment.

Localities.—Gulf of Mexico, North Atlantic, South Atlantic, Panama Bay (stations 2385, 2394, 2550, 2679, 2760, 2784, 2805), 51 to 1,081 fathoms.

NODOSARIA COMMUNIS d'Orbigny.

(Plate 56, fig. 2.)

Slender, tapering, curved; segments numerous, smooth; sutural lines oblique, obvious, little if at all depressed. Length, 2 to 3 mm. ($\frac{1}{12}$ to $\frac{1}{8}$ inch).

Localities.—Off Nantucket Shoals, Gulf of Mexico, off Cape Fear, west coast of Patagonia (stations 2041, 2377, 2679, 2784), 194 to 1,608 fathoms.

NODOSARIA RŒMERI Neugeboren.

(Plate 56, fig. 5.)

Elongate, cylindrical, or slightly tapering, rounded at the base; seg-

NODOSARIA HISPIDA d'Orbigny.

(Plate 57, fig. 1.)

Composed of a linear series of globular segments, each with or without a more or less prolonged tubular neck, arranged usually in a straight line, the whole surface thickly beset with short, mostly tubular spines. Length, about 2.5 mm. ($\frac{1}{10}$ inch).

Locality.—Gulf of Mexico (station 2398), 227 fathoms.

NODOSARIA HISPIDA, variety SUBLINEATA Brady.

(Plate 56, fig. 4.)

Varies from *N. hispida* in that delicate raised lines take the place of spines over a portion of the surface of one or more of the segments.

Locality.—Gulf of Mexico (station 2378), 68 fathoms.

NODOSARIA MUCRONATA Neugeboren.

(Plate 57, fig. 2.)

Elongate, conical, more or less curved, tapering to a point at the aboral end, the final segment also frequently prolonged and conical; sutures oblique and full; surface smooth and even; aperture radiate. Length, about 1.5 mm. ($\frac{1}{18}$ inch).

Localities.—South of Marthas Vineyard and Gulf of Mexico (stations 2550, 2568, 2383), 390, 1,181, and 1,781 fathoms.

NODOSARIA COMATA Batsch.

(Plate 57, fig. 3.)

Ovate or long-oval, tapering and rounded at both ends, composed of a few segments arranged in a straight series; sutures slightly depressed; surface ornamented with numerous longitudinal ridges extending from the extreme point of the initial segment to about the middle of the final one. Length, about 0.75 mm. ($\frac{1}{32}$ inch).

Localities.—Gulf of Mexico, coast of Georgia, off Cape Romain (stations 2352, 2377, 2416, 2627), 210 to 463 fathoms.

NODOSARIA OBLIQUA Linnæus.

(Plate 57, fig. 4.)

Long, slightly curved, tapering, slender, the initial end generally terminating in a spine; segments numerous, the later ones somewhat inflated; sutures more or less depressed; surface ornamented with numerous longitudinal, continuous ridges. Section shows the chambers, cavities, and the communicating passages.

Localities.—Off Atlantic coast of the United States, and the Gulf of Mexico (stations 2264, 2313, 2394, 2530, 2550), 99 to 1,081 fathoms.

NODOSARIA VERTEBRALIS Batsch.

(Plate 57, fig. 5.)

Long, slender, tapering, costate, differing from *N. obliqua* chiefly in that the sutures are not depressed, and the septa are thick and of transparent shell-substance, which contrasts with the white opacity of the body of the segments. Length, about 5 mm. ($\frac{1}{5}$ inch).

Locality.—Gulf of Mexico (stations 2377, 2378, 2399, 2400), 68 to 196 fathoms.

NODOSARIA CATENULATA Brady.

(Plate 58, fig. 2.)

Long, slender, straight or slightly curved, tapering, the initial segment terminating in a short spine; segments numerous; sutures depressed; surface ornament of four equidistant longitudinal ribs, sometimes continuous, sometimes only bridging the sutures and disappearing on the body of the segment. Differs from *N. vertebralis* in its depressed sutures and the limited number of ribs. Length, about 4.5 mm. ($\frac{3}{16}$ inch).

Locality.—Gulf of Mexico (station 2400), 169 fathoms.

NODOSARIA COSTULATA Reuss.

(Plate 58, fig. 1.)

In size and outline the same as *N. pyrula*, but with thicker walls and having the surface ornamented with longitudinal ridges extending sometimes continuously over the whole length of the segments, at other times over only a part of its length.

Locality.—Gulf of Mexico (stations 2377, 2398), 210 and 227 fathoms.

Genus LINGULINA.

Test straight, compressed; aperture typically a narrow fissure.

LINGULINA CARINATA d'Orbigny.

(Plate 58, fig. 3.)

Broad oval or ovate, the margin thin and slightly carinate, smooth; segments four or five, embracing; sutures slightly if at all depressed; aperture a narrow transverse fissure at the end of the final segment. Length, about 1 mm. ($\frac{1}{25}$ inch).

Locality.—Coast of Georgia (station 2416), 276 fathoms.

LINGULINA CARINATA, variety SEMINUDA Hantken.

(Plate 58, fig. 4.)

aperture a transverse slit at the end of the last segment. Section shows the form and arrangement of the chambers. Length, about 1.5 mm. ($\frac{1}{16}$ inch).

Locality.—Gulf of Mexico (stations 2399, 2400), 169 to 170 fathoms.

Genus FRONDICULARIA.

Test compressed or complanate, segments V-shaped, equitant; primordial chamber distinct.

FRONDICULARIA ALATA d'Orbigny.

(Plate 59, fig. 1.)

Triangular or ovate, much compressed, smooth, transparent; commencing usually with a globular chamber, which often bears a projecting spine, the succeeding segments are V-shaped, their arms becoming longer with each additional segment, so that the ends are approximately in line with the initial chamber. Sometimes the earlier segments are irregular, one arm only of the V being developed. Segments numerous; aperture terminal, round, with lateral fissures. Length, 3 mm. ($\frac{1}{8}$ inch), more or less.

Locality.—Gulf of Mexico (stations 2377, 2399), 210 and 198 fathoms.

FRONDICULARIA INÆQUALIS Costa.

(Plate 59, fig. 2.)

Oval or ovate, elongate, smooth; walls thin and fragile; early segments somewhat irregular in form and sequence; the arms of the V-shaped segments short and tapering, seldom reaching the line of the initial chamber. Length, 1.5 mm. ($\frac{1}{20}$ inch), more or less.

Locality.—North Atlantic, off coast of New York (stations 2530, 2584), 956 and 541 fathoms.

Genus MARGINULINA.

Test elongate, curved; segments nearly circular in section; aperture marginal.

MARGINULINA GLABRA d'Orbigny.

(Plate 60, fig. 1.)

Short, stout, smooth, irregularly ovate, slightly curved owing to the planospiral arrangement of the first three segments; the later segments inflated, especially on the inner side of the curve; sutures often indistinct, aperture more or less radiate. Section shows the form and arrangement of the chambers. Length, 1.5 mm. ($\frac{1}{16}$ inch), more or less.

Localities.—North Atlantic (six stations), Straits of Yucatan, Gulf of Mexico (stations 2041, 2234, 2358, 2392, 2570, 2586, 2641, 2677), 60 to 1813 fathoms.

MARGINULINA ENSIS Reuss.

(Plate 59, fig. 3.)

Elongate, subcylindrical, early segments moderately compressed, the first four or five curved so as to form about half a convolution; later segments inflated, arranged in a nearly straight line, with slightly oblique, depressed sutures; surface smooth; walls thin and rather fragile; aperture marginal, tubular, round, with radiating fissures. Length, 2.5 to 4 mm. ($\frac{1}{10}$ to $\frac{1}{6}$ inch).

Locality.—North Atlantic (stations 2242, 2343, 2614), 58 to 168 fathoms.

Genus VAGINULINA.

Test elongate, compressed or complanate; septation oblique; aperture marginal.

VAGINULINA SPINIGERA Brady.

(Plate 60, fig. 3.)

Elongate, compressed, tapering, smooth, bearing at the initial end two or more long stout spines. The earliest two or three segments are spirally arranged; subsequently they are in linear series, with more or less oblique sutural lines. Length of body, 3 mm. ($\frac{1}{8}$ inch), more or less.

Locality.—North Atlantic (stations 2263, 2586), 430 and 328 fathoms.

VAGINULINA LEGUMEN Linnæus.

(Plate 60, fig. 2.)

Elongate, slightly compressed, smooth, of nearly uniform diameter; initial end terminating in a stout marginal spine; oral extremity tapering toward the margin opposite the initial spine; sutures distinct, oblique, not depressed; no surface ornamentation. Length, about 4 mm. ($\frac{1}{6}$ inch).

Locality.—Gulf of Mexico (station 2395), 347 fathoms.

VAGINULINA LINEARIS Montagu.

(Plate 61, fig. 1.)

Elongate, slightly compressed, of nearly uniform diameter, straight or a little curved; segments numerous, the first three or four irregular, the remainder in linear series with the sutures more or less oblique; surface ornamented with many longitudinal or very slightly diagonal ribs. Length, about 2.5 mm. ($\frac{1}{10}$ inch).

Localities.—Off coast of Georgia and Florida (stations 2315, 2416,

CRISTELLARIA TENUIS Bornemann.

(Plate 61, fig. 2.)

A small, elongate, slender, delicate shell, the initial portion compressed; segments numerous, the earliest ones spirally arranged, the others in linear series; walls thin and transparent; sutures near the oral end transverse and more or less depressed; aperture terminal, central. Length, 1.25 mm. ($\frac{1}{20}$ inch), more or less.

Locality.—Atlantic Coast of the United States; station doubtful.

CRISTELLARIA OBTUSATA, variety SUBALATA Brady.

(Plate 61, fig. 3.)

Elongate, slightly compressed and curved, rather broader at the initial than at the oral end; surface smooth; ventral margin rounded, dorsal margin acute and distinctly carinate at the aboral extremity; early segments spiral, later ones linear-oblique; sutures distinct but not depressed. Length, 2.5 to 4 mm. ($\frac{1}{10}$ to $\frac{1}{6}$ inch).

Localities.—Gulf of Mexico, off Cape Fear, and off Santa Lucia, West Indies (stations 2395, 2679, 2754), 347, 782, 880 fathoms.

CRISTELLARIA COMPRESSA d'Orbigny.

(Plate 62, fig. 1.)

More or less elongated, much compressed, broad at the initial end, straight or curved, the early segments plano-spiral with the outer margin more or less broadly carinate, the later segments rectilinear; sutures oblique. Length, 2.5 to 4.7 mm. ($\frac{1}{10}$ to $\frac{3}{16}$ inch).

Localities.—Off Nantucket Shoals, south of Long Island, Gulf of Mexico (stations 2041, 2234, 2385), 730 to 1,608 fathoms.

CRISTELLARIA RENIFORMIS d'Orbigny.

(Plate 62, fig. 2.)

Short, compressed, the peripheral edge sharp and sometimes carinate; segments arranged plano-spirally, except the last two or three, which are applied obliquely, forming a projecting angle, in which the aperture is situated. Length, about 2.5 mm. ($\frac{1}{10}$ inch).

Localities.—North Atlantic (four stations), Gulf of Mexico (stations 2041, 2212, 2568, 2584, 2377, 2385), 210 to 1,780 fathoms.

CRISTELLARIA SCHLOENBACHI Reuss.

(Plate 63, fig. 4.)

Small, elongate, nearly circular in section; spiral portion very short and inconspicuous, the remaining portion consisting of a few diagonal segments with slightly depressed sutures; surface smooth; walls thin and transparent. Length, 0.8 to 1 mm. ($\frac{1}{32}$ to $\frac{1}{25}$ inch). From the Gulf of Mexico (stations 2377, 2700), 210 and 169 fathoms.

CRISTELLARIA VARIABILIS Reuss.

(Plate 63, fig. 1.)

Variable in form, according to stage of development, from circular to elongate, compressed; margins generally carinate; young specimens consist of the spiral segments only; older ones have two or three oblique segments added; walls thin and transparent. Length, about 0.4 mm. ($\frac{1}{64}$ inch).

Localities.—Caribbean Sea, North Atlantic, Gulf of Mexico (stations 2144, 2263, 2584, 2378, 2394, 2398), 68 to 896 fathoms.

CRISTELLARIA CREPIDULA Fichtel and Moll.

(Plate 63, fig. 2.)

Elongate or elongate-oval, compressed, smooth, the early spiral arrangement of segments soon changing into the linear-oblique; peripheral margin rounded; sutures slightly depressed. Length, 0.8 to 3 mm. ($\frac{1}{32}$ to $\frac{1}{8}$ inch).

Localities.—Off coast of North Carolina, Georgia, Florida, and west coast of Cuba (stations 2614, 2313, 2416, 2641, 2352), 60 to 463 fathoms.

CRISTELLARIA ACUTAURICULARIS Fichtel and Moll.

(Plate 63, fig. 5.)

Small, ovoid, thick, smooth, with rounded margins; early segments small, closely spiral; later segments increasing rapidly in length and thickness, becoming oblique instead of radial, and somewhat inflated. Length, about 0.6 mm. ($\frac{1}{40}$ inch).

Localities.—Off Carysfort Light, Florida, and off the coast of South Carolina (stations 2641, 2313), 60 to 99 fathoms.

CRISTELLARIA LATIFRONS Brady.

(Plate 63, fig. 3.)

Elongate, triangular in transverse section, tapering toward each end; dorsal angle acute and carinate; ventral face broad, flat, or rounded, with acute or rounded marginal angles; early segments closely spiral, later ones growing rapidly longer and more obliquely set, the final one erect and extending nearly the whole length of the shell. Length, 1 mm. or less ($\frac{1}{25}$ inch).

Localities.—Off Carysfort Light, Florida, and Gulf of Mexico (stations 2641, 2377), 60 to 210 fathoms.

ment comparatively flat and triangular; surface smooth; aperture at the dorsal angle. Length, about 2 mm. ($\frac{1}{12}$ inch).

Localities.—Off coast of Georgia and Gulf of Mexico (stations 2415, 2399), 440 and 196 fathoms.

CRISTELLARIA GIBBA d'Orbigny.

(Plate 64, fig. 1.)

Sublenticular, equally biconvex, smooth, characterized by the somewhat inflated and protuberant final segment, and its contracted septal face. Diameter, about 1 mm. ($\frac{1}{25}$ inch).

Localities.—North Atlantic (three stations), Gulf of Mexico, coast of Yucatan (stations 2243, 2312, 2415, 2379, 2400, 2354), 63 to 1,467 fathoms.

CRISTELLARIA ARTICULATA Reuss.

(Plate 64, fig. 2.)

Test rotaliform, or sometimes with the last few segments more or less evolute; margin rounded or subcarinate; segments slightly inflated; aperture radiate, in the protuberant end of the last segment.

Localities.—Gulf of Mexico and off the coast of Georgia (stations 2399, 2400, 2416), 169 to 276 fathoms.

CRISTELLARIA ROTULATA Lamarck.

(Plate 64, fig. 4.)

Lenticular, biconvex, smooth; margin sharp, but not carinate; formed of about three convolutions, the last entirely inclosing the others; walls thick and strong. Section shows well the form and arrangement of the chambers and their apertures and the structure of the shell. Diameter, 1.5 to 2.5 mm. ($\frac{1}{16}$ to $\frac{1}{10}$ inch).

Localities.—Caribbean Sea, North Atlantic, and Gulf of Mexico (stations 2150, 2415, 2399), 196 to 440 fathoms.

CRISTELLARIA VORTEX Fichtel and Moll.

(Plate 65, fig. 1.)

Lenticular, biconvex, smooth, with a sharp noncarinate margin; distinguished by the long helicoid curve of the sutures marking the outline of the chambers. Diameter, about 1 mm. ($\frac{1}{25}$ inch).

Localities.—North Atlantic and Caribbean Sea (stations 2416, 2357), 276 and 130 fathoms.

CRISTELLARIA ORBICULARIS d'Orbigny.

(Plate 64, fig. 3.)

Form of the shell and the shape and arrangement of the chambers same as in *C. vortex.* Differs only in having the margin extended into a distinct wing or keel.

Locality.—Gulf of Mexico (stations 2377, 2400), 210 and 169 fathoms.

CRISTELLARIA CULTRATA Montfort.

(Plate 65, fig. 2.)

A lenticular, biconvex, smooth shell, in all general characters like *C. rotulata* except the peripheral margin, which in this species is extended into a thin, broad wing or keel. Diameter, 2 mm. ($\frac{1}{12}$ inch), more or less.

Locality.—Gulf of Mexico (stations 2399, 2400), 196 and 169 fathoms.

CRISTELLARIA CALCAR Linnæus.

(Plate 66, fig. 1.)

Lenticular, biconvex, smooth, carinate, in some instances with a broad keel notched and spinous at the edge, in other cases with a narrow keel and long, slender, radiating spines. Size variable; the large specimens generally have the broad keel and the small ones the long spines.

Localities.—Off the coast of the Carolinas and in the Gulf of Mexico (stations 2312, 2313, 2679, 2377, 2400), 88 to 782 fathoms.

CRISTELLARIA ECHINATA d'Orbigny.

(Plate 66, fig. 2.)

Test lenticular; margin either rounded or keeled and projected into more or less numerous radiating processes; sutures limbate and beaded. Diameter, 1.25 to 2.50 mm. ($\frac{1}{20}$ to $\frac{1}{10}$ inch).

Locality.—Gulf of Mexico (stations 2377, 2399, 2400), 169 to 210 fathoms.

CRISTELLARIA ACULEATA d'Orbigny.

(Plate 66, fig. 3.)

Elongate, moderately compressed; early segments planospiral, later ones rectilinear or curved; sutures oblique and conspicuously marked by rounded tubercles or short, stout spines; general surface, especially of the earlier segments, often tuberculated or spinous, peripheral edge sometimes finished with several long, slender spines.

Locality.—Gulf of Mexico (stations 2377, 2399), 210 and 196 fathoms.

CRISTELLARIA LIMBATA, new species.

(Plate 67, fig. 1.)

Elongate, evolute, slightly compressed, resembling *C. aculeata* in

Subfamily POLYMORPHININÆ.

Segments arranged spirally or irregularly around the long axis; rarely biserial and alternate.

Genus POLYMORPHINA.

Segments bi- or tri- serial or irregularly spiral; aperture radiate.

POLYMORPHINA SORORIA Reuss, variety FISTULOSA.

(Plate 67, fig. 2.)

Body ovate, smooth, nearly symmetrical, composed of four or five elongated segments, arranged spirally. Upon the symmetrical body is set a final segment, irregularly globular, rough, bearing numerous slender, tubular, radiating projections with a round aperture at the end of each.

Localities.—North Atlantic, off coast of Brazil, Gulf of Mexico (stations 2221, 2568, 2763), 671 to 1,781 fathoms.

POLYMORPHINA COMPRESSA d'Orbigny.

(Plate 67, fig. 3.)

Irregularly oval, compressed, smooth, margins rounded; composed of four to eight segments arranged in two alternating series; aperture terminal, radiate; sutures more or less depressed. Length, 0.8 to 1.6 mm. ($\frac{1}{32}$ to $\frac{1}{16}$ inch).

Localities.—Off Atlantic Coast of the Southern United States (stations 2312, 2313, 2415, 2416, 2614), 88 to 440 fathoms.

POLYMORPHINA ELEGANTISIMA Parker and Jones.

(Plate 67, fig. 4.)

Ovate or pyriform, compressed unequally on two sides; margins rounded, surface smooth, segments long and arched, arranged biserially, but the alternation inequilateral; aperture terminal, radiate. Length, 1 mm. ($\frac{1}{25}$ inch) or less.

POLYMORPHINA OBLONGA d'Orbigny.

(Plate 67, fig. 5.)

Oval, elongate, more or less compressed, composed of about six oblong, inflated segments, unsymmetrically arranged and united by depressed sutures.

Localities.—Off the coast of Georgia and North Carolina (stations 2416, 2614), 276 and 168 fathoms.

POLYMORPHINA COMMUNIS d'Orbigny.

(Plate 67, fig. 6.)

Ovate, not compressed; visible segments three or four, oval, inflated, symmetrically arranged; sutures rather indistinct, not depressed. Length, 0.8 to 0.6 mm. ($\frac{1}{30}$ to $\frac{1}{40}$ inch).

Localities.—Off coast of Georgia and off Unalaska (stations 2416, 2842).

Genus UVIGERINA.

Segments arranged in a more or less regular spire around the long axis of the shell, rarely biserial. Aperture simple, usually surrounded by a phialine lip; often forming a prolonged terminal tube.

UVIGERINA TENUISTRIATA Reuss.

(Plate 68, fig. 1.)

Oval, elongate; sutures not well marked; arrangement of segments obscure; surface ornamented with numerous very fine longitudinal striæ; aperture tubular, with a phialine lip, the tube sometimes bearing two or three rings of shell substance. Length, about 0.6 mm. ($\frac{1}{40}$ inch).

Locality.—Off Carysfort Light, Florida (station 2641), 60 fathoms.

UVIGERINA PYGMÆA d'Orbigny.

(Plate 68, fig. 2.)

Oval, more or less elongated, symmetrical; surface rough with thin, prominent, interrupted costæ; aperture tubular with a phialine lip. The principal feature distinguishing this species from *U. tenuistriata* is the prominence of the costæ.

Locality.—Off Cape Fear (station 2679), 782 fathoms.

UVIGERINA ANGULOSA Williamson.

(Plate 68, fig. 3.)

Small, elongate, compressed on three sides, the sides nearly equal, the angles sharp, surface roughened with more or less prominent costæ. Length, about 0.4 mm. ($\frac{1}{60}$ inch).

Localities.—Exuma Sound and Panama Bay (stations 2530, 2805), 956 and 51 fathoms.

UVIGERINA ASPERULA Czjzek.

(Plate 68, fig. 4.)

Oval or ovate, more or less elongated, rounded at the initial end, the surface roughened with short spines, sometimes set in rows and tending to run together into short costæ, at other times, especially on the terminal segment, irregularly and closely distributed; aperture phialine on a tubular neck. Length, about 0.5 mm. ($\frac{1}{50}$ inch).

Locality.—Off the coast of Brazil (station 2760), 1,019 fathoms.

neck with a phialine lip; surface, bristly-spiny. Length, about 0.6 mm. ($\frac{1}{40}$ inch).

Locality.—Off the Brazil coast (station 2760), 1,019 fathoms.

Subfamily RAMULININÆ.

Test irregular, branching.

Genus RAMULINA.

Test branching, composed of pyriform chambers connected by long stoloniferous tubes.

RAMULINA GLOBULIFERA Brady.

(Plate 68, fig. 6.)

Segments few, subglobular, united by long stoloniferous tubes, and each segment provided with numerous radiating tubulures; walls hyaline; surface bristly with sparsely set fine and short spines.

Locality.—Gulf of Mexico (station 2377,) 210 fathoms.

RAMULINA PROTEIFORMIS, new species.

(Plate 68, fig. 7.)

Test calcareous, extremely thin and fragile, very finely perforated; surface smooth; in form very irregular and variable, sometimes branching, sometimes with more or less numerous short digital processes, imperfectly segmented, the segments inflated into a great variety of shapes. The figures show only a few of the myriad forms assumed by this delicate foraminifer.

Locality.—Gulf of Mexico (stations 2352 and 2377), 463 and 210 fathoms.

Family VIII. GLOBIGERINIDÆ.

Test free, calcareous, perforate; chambers few, inflated, arranged spirally; aperture single or multiple, conspicuous.

Genus GLOBIGERINA.

Test coarsely perforate; trochoid, rotaliform, or symmetrically planospiral.

GLOBIGERINA BULLOIDES d'Orbigny.

(Plate 69, fig. 2.)

Subglobular, the adult shell composed of about seven nearly spherical segments, arranged spirally so that all are visible on the upper side, and three or four on the lower side; aperture of each chamber opens into a common umbilical vestibule; surface more or less rough; walls hyaline, finely and distinctly perforated. Diameter, 0.6 mm. ($\frac{1}{40}$ inch) or less.

Locality.—Coast of Yucatan (station 2358), 222 fathoms. Found in almost every part of the ocean.

GLOBIGERINA INFLATA d'Orbigny.

(Plate 69, fig. 3.)

Subglobular, flattened on the superior face; segments rather numerous, four in the final convolution; sutures depressed; aperture a large arched gaping orifice on the face of the final segment. Diameter, about 0.5 mm. ($\frac{1}{50}$ inch). Found in almost every sea.

Localities.—North Atlantic and the Gulf of Mexico (stations 2204, 2372), 728 and 27 fathoms.

GLOBIGERINA DUBIA Egger.

(Plate 69, fig. 4.)

Subglobular, slightly compressed, segments relatively numerous, arranged spirally in about three convolutions, all the segments visible on the upper face, five or six on the lower; umbilical vestibule central, with which all the chambers directly connect; surface rough; walls finely perforated. Diameter, about 0.6 mm. ($\frac{1}{40}$ inch).

Locality.—Panama Bay. Species widely distributed.

GLOBIGERINA RUBRA d'Orbigny.

(Plate 69, fig. 5.)

Shell composed of nearly globular segments, arranged in a spire of about three convolutions with three segments in each whorl; apertures, a single, large, arched orifice in the face of the final segment and one or two rounded openings on the superior face of several of the chambers near the sutures; surface rough; walls finely perforated; color pink. Diameter, about 0.5 mm. ($\frac{1}{50}$ inch).

Localities.—Widely distributed. Specimens taken off the Windward Islands and the coast of Brazil (stations 2751, 2760), 687 and 1,019 fathoms.

GLOBIGERINA CONGLOBATA Brady.

(Plate 69, fig. 6.)

Subglobular, the early segments comparatively small and compact, the last three large and inflated, the final one resting like a cap upon one side of the shell; surface rough, originally bristly-spiny, as shown by the unbroken spines in the aperture; principal aperture broad and arched at the margin of the last segment, other small orifices in the sutural depressions on the upper side of the shell; walls thick and profusely perforated. Diameter, about 0.8 mm. ($\frac{1}{30}$ inch).

Localities.—Widely distributed. Specimens from Windward Islands and coast of Brazil (stations 2751, 2760), 687 and 1,019 fathoms.

inflated into various and irregular forms, the peripheral extremity often bearing several short digital outgrowths; apertures multiple, large, often five visible on the superior face; walls conspicuously perforated. Diameter, 1 mm. ($\frac{1}{25}$ inch), more or less.

Localities.—Found in tropical and subtropical latitudes. Specimens from the same stations as the two preceding.

GLOBIGERINA DIGITATA Brady.

(Plate 70, fig. 2.)

Early segments spiral, regular, same as *G. bulloides;* last three segments of the final convolution elongated and rounded at the ends like the fingers of a glove, spreading radially.

Locality.—A single specimen from the Gulf of Mexico (station 2377), 210 fathoms.

GLOBIGERINA ÆQUILATERALIS Brady.

(Plate 70, fig. 3.)

Segments subglobular, increasing rather rapidly in size, arranged in a flat coil of about one convolution and half another, all the segments being equally visible on both sides; aperture a large arched opening on the inner face of each segment; walls conspicuously perforated; surface rough with the short stumps of broken spines. Diameter, about 0.8 mm. ($\frac{1}{30}$ inch).

Locality.—Specimens dredged off the Windward Islands (station 2751), 687 fathoms.

Genus ORBULINA.

Test having the form of a single spherical chamber with two sorts of perforations, large and small.

ORBULINA UNIVERSA d'Orbigny.

(Plate 69, fig. 1.)

Typically in the form of a perfect sphere with thin walls inclosing a single chamber; occasionally two or three chambered shells are found; walls sometimes laminated, profusely perforated with both very fine and comparatively large orifices. No general aperture. Diameter, about 0.8 mm. ($\frac{1}{30}$ inch).

Localities.—The most common of all the species of foraminifera. Found in every sea.

Genus HASTIGERINA.

Test regularly nautiloid, involute; shell wall thin, finely perforated; armed with long serrate spines. Aperture a large crescentiform opening at the base of the last chamber.

HASTIGERINA PELAGICA d'Orbigny.

(Plate 70, fig. 4.)

Subglobular, compressed equally on both sides, umbilici depressed; composed of inflated segments rapidly increasing in size, arranged in a planospiral series of about two convolutions, the last convolution entirely including the others; walls thin; sutures depressed; surface roughish with the stumps of broken spines; aperture a large arched opening at the inner margin of the last segment. Diameter, about 0.8 mm. ($\frac{1}{30}$ inch).

Locality.—Specimens exhibited are worn bottom shells collected in the Gulf of Mexico (station 2377), 210 fathoms.

Genus PULLENIA.

Test regularly or obliquely nautiloid and involute; segments only slightly ventricose; shell wall very finely perforated; aperture a long, curved slit close to the line of union of the last segment with the previous convolution.

PULLENIA QUINQUELOBA Reuss.

(Plate 70, fig. 5.)

Biconvex, bilaterally symmetrical, round, peripheral edge thick and rounded, final convolution consisting of about five segments wholly concealing the previous convolutions; surface smooth; sutures sometimes depressed, sometimes obscure; aperture a long, narrow, curved slit at the inner margin of the last segment. Diameter, about 0.6 mm. ($\frac{1}{40}$ inch).

Localities.—Widely distributed; specimens from the North Atlantic (three stations) and the Gulf of Mexico (stations 2115, 2204, 2584, 2352), 463 to 843 fathoms.

PULLENIA OBLIQUILOBULATA Parker and Jones.

(Plate 70, fig. 6.)

Subglobular, slightly compressed, inequilateral, obliquely nautiloid; surface smooth; walls thick and finely but conspicuously perforated; aperture a crescentic opening on the inner margin of the last segment, generally somewhat obliquely placed. Diameter, about 0.8 mm. ($\frac{1}{30}$ inch).

Locality.—Off the Windward Islands, West Indies (station 2751), 687 fathoms.

SPHÆROIDINA BULLOIDES d'Orbigny.

(Plate 71, fig. 1.)

Nearly spherical, smooth, composed of comparatively few segments arranged in an approximately symmetrical spire; sutures slightly depressed; walls minutely and indistinctly perforated; aperture semicircular or crescentic, sometimes with a valvular lip, at the inner margin of the last segment. Diameter, about 1 mm. ($\frac{1}{25}$ inch).

Localities.—Widely distributed; specimens from North Atlantic, Gulf of Mexico, and South Atlantic (stations 2530, 2383, 2760), 956 to 1,181 fathoms.

SPHÆROIDINA DEHISCENS Parker and Jones.

(Plate 71, fig. 2.)

Subglobular; segments arranged as in *S. bulloides;* sutures at the bottom of wide and deep irregular fissures; walls thick and conspicuously perforated; aperture an arched opening into the deep fissure at the base of the last segment. Diameter, about 1 mm. ($\frac{1}{25}$ inch).

Localities.—Caribbean Sea, Gulf of Mexico, and off Windward Islands (stations 2150, 2358, 2399, 2751), 196 to 687 fathoms.

Genus CANDEINA.

Test trochoid; segments inflated; shell-walls thin, finely perforated; aperture consisting of rows of pores along the septal depressions.

CANDEINA NITIDA d'Orbigny.

(Plate 71, fig. 3.)

Contour irregular, subconical; segments twelve to fifteen, subspherical, smooth, regularly increasing in size, arranged in an elongated spiral; sutures deeply depressed, walls thin and very minutely perforated; aperture a series of pores rather closely set in the sutures uniting the segments. Diameter, about 0.5 mm. ($\frac{1}{50}$ inch).

Locality.—Specimens taken near the Windward Islands (station 2751), 687 fathoms.

Family IX. ROTALIDÆ.

Test calcareous, perforated; free or adherent. Typically spiral and "rotaliform;" that is to say, coiled in such a manner that all the segments are visible on the superior surface, those of the last convolution only on the inferior or apertural side, sometimes one face being more convex, sometimes the other.

Subfamily SPIRILLININÆ.

Test spiral, nonseptate.

Genus SPIRILLINA.

Test a complanate, planospiral, nonseptate tube; free or attached.

SPIRILLINA VIVIPARA Ehrenberg.

(Plate 71, fig. 4.)

A circular, double concave disk, formed by a single tube closely coiled in one plane; tube undivided, conspicuously perforated by a single row of pores; sutures thick, but not raised; aperture, the open end of the unconstricted tube. Diameter, 0.75 mm. ($\frac{1}{35}$ inch) or less.

Localities.—Not recorded.

SPIRILLINA LIMBATA Brady.

(Plate 71, fig. 5.)

Circular, concave on both sides, composed of numerous regular coils of a flattened tube; peripheral edge square; sutural line marked by a raised ridge of shell substance; general surface smooth; perforations very indistinct. Diameter, about 0.8 mm. ($\frac{1}{30}$ inch).

Locality.—Not recorded.

SPIRILLINA OBCONICA Brady.

(Plate 71, fig. 6.)

Circular, deeply concave on one side, moderately convex on the other; peripheral edge rounded; sutures deeply depressed on the concave face, flush on the other; convolutions eight or ten; perforations on the concave face only, at the summit of minute bead-like prominences arranged in a single row along the sutural side of the tube; tube slightly constricted at regular intervals alternating with the perforations. Diameter, 0.8 to 1.2 mm. ($\frac{1}{30}$ to $\frac{1}{20}$ inch).

Locality.—Not recorded.

Subfamily ROTALINÆ.

Test spiral, rotaliform, rarely evolute, very rarely irregular or acervuline.

Genus CYMBALOPORA.

Test more or less trochoid or complanate. Segments of the trochoid forms spiral at the apex, subsequently arranged concentrically around a deep umbilical vestibule with which each chamber communicates by a neck. Complanate forms with rows of pores along the septal depressions of the inferior surface.

CYMBALOPORA POEYI d'Orbigny.

(Plate 72, fig. 1.)

porous; aperture of each chamber opens into the central vestibule. Diameter, about 0.75 mm. ($\frac{1}{33}$ inch).

Locality.—Off the west coast of Cuba (station 2352), 463 fathoms.

Genus DISCORBINA.

Test free or adherent, rotaliform; plano-convex or trochoid; rarely complanate; aperture an arched slit, often protected by an umbilical flap, the flaps sometimes forming a whorl of subsidary chambers.

DISCORBINA GLOBULARIS d'Orbigny.

(Plate 72, fig. 2.)

Discoidal, thick, the superior face quite convex, the inferior only slightly so; segments somewhat inflated, finely perforated, hyaline, all visible superiorly, only the last convolution inferiorly; sutures a little depressed; aperture large and irregular at the umbilical margin of the last segment. Diameter, about 0.8 mm. ($\frac{1}{30}$ inch).

Locality.—Off Carysfort Light, Florida (station 2641), 60 fathoms.

DISCORBINA ROSACEA d'Orbigny.

(Plate 72, fig. 3.)

Contour lenticular, plano-convex, peripheral margin rounded; composed of about three convolutions of six segments each; surface smooth and polished; sutures distinct but not depressed; color, pale brown; aperture a narrow arched slit at the umbilical margin of the final segment. Diameter, about 0.4 mm. ($\frac{1}{60}$ inch).

Locality.—Coast of Alaska, station unknown.

DISCORBINA BERTHELOTI d'Orbigny.

(Plate 72, fig. 4.)

Discoidal, thin, plano-convex; the superior convex face somewhat flattened at the center, peripheral margin sharp; outlines of the segments very distinct; sutures a little depressed, and thickened with transparent shell-substance; later segments moderately inflated; walls finely but distinctly perforated. Diameter, about 0.4 mm. ($\frac{1}{60}$ inch).

Localities.—North Atlantic and Gulf of Mexico (stations 2212, 2313, 2352), 79 to 463 fathoms.

DISCORBINA BICONCAVA Parker and Jones.

(Plate 72, fig. 5.)

Circular, flattened on both faces; peripheral margin square or slightly concave; coarsely perforated; sutures on the superior face between the earlier segments raised into prominent, thin, square-edged, wavy ridges; on the inferior face only slightly limbate. Diameter, about 0.4 mm. ($\frac{1}{60}$ inch).

Locality.—Gulf of Mexico (station 2400), 169 fathoms.

Genus PLANORBULINA.

Test normally adherent; compressed or complanate segments very numerous, commencing growth on a spiral plan, subsequently becoming more or less cyclical; lipped apertures of the individual segments opening externally at the periphery.

PLANORBULINA ACERVALIS Brady.

(Plate 72, fig. 7.)

Discoidal, thin, the attached side flat and smooth, the inferior face roughened by the projection of numerous irregular inflated segments over the whole surface; walls coarsely porous; apertures peripheral. Diameter, 1.5 to 2.5 mm. ($\frac{1}{16}$ to $\frac{1}{10}$ inch).

***Locality.*—Gulf of Mexico (station 2399), 190 fathoms.**

PLANORBULINA MEDITERRANENSIS d'Orbigny.

(Plate 72, fig. 6.)

A thin, flat, nearly circular shell, when living usually attached to some foreign body, composed of numerous segments arranged in a single layer more or less distinctly spiral; attached surface nearly flat, opposite surface lobulated; periphery irregular; segments inflated, slightly embracing, very conspicuously and profusely perforated; sutures depressed; apertures at the extremity of each segment, simple, with a raised lip. Diameter, about 1 mm. ($\frac{1}{25}$ inch).

Locality.—A single specimen obtained in the Gulf of Mexico (station 2377), 210 fathoms.

Genus PULVINULINA.

Test rotaliform, superior side usually thickest; shell finely porous; segments fewer than in other rotalinæ; aperture typically a large slit at the base of the umbilical margin of the last segment.

PULVINULINA REPANDA Fichtel and Moll.

(Plate 72, fig. 8.)

Lenticular, about equally convex on both faces; peripheral margin subacute, limbate; sutures broad, conspicuous by reason of their glassy clearness, limbate on both faces; umbilicus filled smoothly with hyaline shell substance; aperture as usual.

Locality.—Arrowsmith Bank, coast of Yucatan (station 2354), 130 fathoms.

PULVINULINA PUNCTULATA d'Orbigny.

cus narrowed by promontories of exogenous deposit. Diameter, 1 to 1.5 mm. ($\frac{1}{25}$ to $\frac{1}{16}$ inch).

Locality.—Coast of Georgia (stations 2415, 2416), 440 and 276 fathoms.

PULVINULINA AURICULA Fichtel and Moll.

(Plate 73, fig. 2.)

Long oval in contour, biconvex, the convexity of the two sides about equal, the earlier segments closely coiled, the later ones rapidly increasing in size, especially in length; walls thin, transparent, and finely perforated; sutures distinct, but not depressed or thickened; margin sharp, but not carinate. Length, 0.5 to 1 mm. ($\frac{1}{50}$ to $\frac{1}{25}$ inch).

Locality.—Gulf of Mexico (stations 2400, 2641), 169 and 60 fathoms.

PULVINULINA MENARDII d'Orbigny.

(Plate 73, fig. 3.)

Contour subcircular, much flattened, composed of about two convolutions of slightly inflated segments, all visible on the upper side, the six forming the final whorl visible on the lower side; margin thin, slightly lobed, and with a narrow keel; sutures broad, but not raised, slightly depressed on the superior side; aperture a wide slit at the inner margin of the last segment, often with a protruding under lip. Diameter, about 1.25 mm. ($\frac{1}{20}$ inch).

Localities.—A very common and widely distributed species. Specimens collected off Windward Islands, West Indies (station 2751), 687 fathoms.

PULVINULINA MENARDII, variety FIMBRIATA Brady.

(Plate 73, fig. 4.)

Has the same general characters as the type, but is smaller, and is distinguished by the fringed peripheral border produced by the development of numerous short spinous processes upon the normal narrow keel.

Locality.—Coast of Brazil (station 2760), 1,019 fathoms.

PULVINULINA TUMIDA Brady.

(Plate 73, fig. 5.)

Like *P. menardii*, except that the segments are more inflated, making a thicker shell, highly convex on both faces; margin not carinate; sutures slightly, if at all, depressed below.

Localities.—Off coast of Yucatan and coast of Georgia (stations 2354, 2416), 130 and 276 fathoms.

PULVINULINA CRASSA d'Orbigny.

(Plate 74, fig. 1.)

Superior face flat, showing all the convolutions; inferior face highly conical, composed of the final convolution only; umbilicus depressed; segments somewhat inflated; walls hyaline, profusely and finely perforated; exteriorly rough; aperture a long fissure with a raised lip at the inner margin of the final segment. Section shows chambers of the final convolution. Diameter, about 0.6 mm. ($\frac{1}{40}$ inch).

Locality.—Not recorded.

PULVINULINA MICHELIANA d'Orbigny.

(Plate 74, fig. 2.)

Subconical, the superior face forming the base of the cone, being flat with an angular margin; the inferior face being conical, deeply excavated at the top; segments, about ten, elongated, projecting in a ridge around the umbilicus; sutures not depressed; aperture a long narrow slit at the inner margin of the last segment. Transverse section close to the superior surface has opened all but one of the ten chambers. Diameter, about 0.8 mm. ($\frac{1}{32}$ inch).

Localities.—Species widely distributed geographically. Specimens from the Gulf of Mexico (station 2377), 210 fathoms.

PULVINULINA UMBONATA Reuss.

(Plate 74, fig. 4.)

Small, biconvex, with the greatest convexity on the lower face; umbilici not depressed; margin rounded; segments rather numerous, in about three narrow convolutions; sutures straight, radial, smooth. Diameter, about 0.75 mm. ($\frac{1}{33}$ inch).

Locality.—Off coast of Oregon (station 3080).

PULVINULINA PAUPERATA Parker and Jones.

(Plate 74, fig. 3.)

Thin, flat, and transparent, composed of fifteen to twenty or more slightly inflated segments, arranged in about two planospiral convolutions, all the segments being visible on both sides; margin extended into a broad, thin wing of clear shell-substance entirely surrounding the final convolution. Diameter, about 1.5 mm. ($\frac{1}{16}$ inch), often much greater.

Locality.—Specimens from the Gulf of Mexico (stations 2385, 2395), 730 and 347 fathoms.

PULVINULINA KARSTENI Reuss.

(Plate 74, fig. 5.)

Lenticular, about equally convex on both faces, smooth and regular, with a blunt angular peripheral margin, composed of about three con-

PULVINULINA ELEGANS d'Orbigny.

(Plate 75, fig. 1.)

Lenticular, about equally convex on the two sides, smooth; peripheral margin rounded; sutures well marked but not elevated or depressed; walls clear, transparent, and beautifully marked by opaque-white, broad, wavy lines and irregular dots; aperture at the inner margin of the final segment, a second aperture is found in most specimens as a linear slit just beneath the peripheral margin of the last segment. Diameter, about 1.5 mm. ($\frac{1}{16}$ inch).

Localities.—Gulf of Mexico, North Atlantic, and Panama Bay (stations 2352, 2394, 2570, 2805), 51 to 1,813 fathoms.

PULVINULINA PARTSCHIANA d'Orbigny.

(Plate 75, fig. 3.)

Differs from *P. elegans* in its smaller size, the tendency to limbation of the sutures, and especially in the absence of the variegated markings which give the specific name to the former species. Diameter, about 0.75 mm. ($\frac{1}{32}$ inch).

Locality.—Gulf of Mexico (station 2394), 420 fathoms.

Genus ROTALIA.

Test rotaliform, shell-wall very finely porous; exogenous deposit either in the form of embossed septal lines or of granulation of the sutures near the umbilicus. Aperture a neatly arched slit, nearly median.

ROTALIA BECCARII Linnæus.

(Plate 75, fig. 2.)

Double-convex, with convexity greatest on the inferior face; margin rounded and slightly lobulated; segments numerous, arranged in about four convolutions, only the last visible on the under side; upper surface smooth; septa on inferior face more or less raised and granular, in some cases double, with a deep fissure between the layers; umbilicus sometimes excavated, sometimes filled with clear shell-substance; walls thick and strong. Diameter, about 0.8 mm. ($\frac{1}{30}$ inch).

Locality.—Not recorded.

ROTALIA ORBICULARIS d'Orbigny.

(Plate 75, fig. 5.)

Superior face flat or slightly convex, inferior face moderately and regularly convex; umbilicus scarcely if at all depressed; peripheral margin rounded; walls finely porous; surface smooth, without ornamentation; segments numerous, twelve or more in the final convolution; sutures conspicuous because of the thickening of the septal walls; orifice regular. Diameter, about 0.8 mm. ($\frac{1}{30}$ inch).

Locality.—Coast of Oregon (station 3080), 93 fathoms.

ROTALIA SOLDANII d'Orbigny.

(Plate 75, fig. 4.)

Superior face flat and smooth; inferior face highly convex; umbilicus deeply excavated; peripheral margin thick and well rounded; walls very finely perforated, surface smooth except the granular umbilicus; face of the final segment broad and flat. Diameter, about 1 mm. ($\frac{1}{25}$ inch).

Localities.—A deep-water species, widely distributed. Specimens from North Atlantic, Gulf of Mexico, and North Pacific (stations 2115, 2228, 2550, 2568, 2570, 2385, 2394, 3080).

ROTALIA SCHROETERIANA Parker and Jones.

(Plate 76, fig. 1.)

A large, strong, symmetrical shell, slightly convex on the upper face, highly convex below; sutures broad and conspicuously marked on both faces by numerous prominent beads of clear shell-substance; umbilicus filled with a dense irregular mass of shell. Section near the superior face has opened all the chambers; cross section shows the umbilical mass of shell-substance. Diameter, about 1.5 mm. ($\frac{1}{16}$ inch).

Locality.—Not recorded.

ROTALIA PAPILLOSA Brady.

(Plate 76, fig. 2.)

Test lenticular, nearly equally convex on the two faces; segments clearly defined on both faces by thick septa of transparent shell-substance more or less regularly penetrated by round apertures sometimes running into short fissures. Diameter, about 1 mm. ($\frac{1}{25}$ inch).

Locality.—Not recorded.

ROTALIA PULCHELLA d'Orbigny.

(Plate 76, fig. 3.)

Small, much compressed on both faces, composed of numerous somewhat inflated segments arranged in three or four convolutions, only the last convolution visible on the underside; sutures raised in narrow, sometimes interrupted ridges. Projecting radially from the margin are three or four long slender spines, equaling or exceeding in length, when unbroken, the greatest diameter of the test. Diameter, about 0.4 mm. ($\frac{1}{60}$ inch).

Locality.—Not recorded.

Genus TRUNCATULINA

TRUNCATULINA LOBATULA Walker and Jacob.

(Plate 76, fig. 4.)

Planoconvex, the convexity on the inferior face; peripheral margin rounded; segments rather numerous, only the final convolution visible below; walls stout and coarsely porous; sutures thickened with clear shell-substance and more or less limbate near the umbilici; aperture a long fissure at the upper and inner margin of the last segments. Diameter, from 0.8 to 1.2 mm ($\frac{1}{30}$ to $\frac{1}{20}$ inch).

Locality.—Bahia, Brazil (station 2760), 1,019 fathoms.

TRUNCATULINA WUELLERSTORFI Schwager.

(Plate 77, fig. 1.)

Outline circular, much compressed, inferior face moderately convex, superior face flat or slightly concave, peripheral margin sharp; composed of numerous narrow curved segments arranged in about three convolutions; walls coarsely porous; aperture regular. Diameter, about 1.25 mm. ($\frac{1}{20}$ inch).

Localities.—Gulf of Mexico, North Atlantic, and Panama Bay (stations 2150, 2370, 2392, 2570, 2565, 2750, 2805), 25 to 2,069 fathoms.

TRUNCATULINA UNGERIANA d'Orbigny.

(Plate 77, fig. 2.)

Nearly equally convex on the two surfaces, peripheral margin thin. Differs from *T. wuellerstorfi* in that the superior face is convex, the segments shorter and less curved, and the walls less coarsely porous.

Localities.—Gulf of Mexico and coast of Brazil (stations 2078, 2393, 2400, 2760), 169 to 1,019 fathoms.

TRUNCATULINA AKNERIANA d'Orbigny.

(Plate 77, fig. 5.)

Circular, compressed, superior surface flat, inferior convex at the margin, flat toward the center; margin rounded; a more or less deep and extended fissure on the superior face between the last convolution and the preceding one. Section shows the chambers of the last convolution and a portion of the next. Diameter, about 1.25 mm. ($\frac{1}{20}$ inch).

Localities.—Gulf of Mexico and coast of Brazil (stations 2377, 2394, 2398, 2760), 210 to 1,019 fathoms.

TRUNCATULINA ROBERTSONIANA Brady.

(Plate 77, fig. 3.)

Superior surface nearly flat, inferior convex, but flattened toward the center; margin thick and rounded; walls quite transparent, showing clearly the convolutions and the outlines of the numerous segments; all the convolutions visible on the upper face, on the lower face the final convolution leaves exposed some of the earlier segments; walls coarsely porous; color often a more or less deep shade of brown. Diameter, about 0.7 mm. ($\frac{1}{35}$ inch).

Localities.—North Atlantic, Gulf of Mexico, Caribbean Sea, coast of Brazil (stations 2568, 2352, 2392, 2394, 2760), 463 to 1,781 fathoms.

TRUNCATULINA TENERA Brady.

(Plate 77, fig. 4.)

Small, discoidal, inferior face the more convex; peripheral margin acute and slightly lobulated; visible segments on the inferior face six or seven; convolutions about three of nearly equal width; sutures slightly depressed, straight and radial; aperture a short curved fissure with thickened lip, at the inner margin of the final segment. Diameter, about 0.5 mm. ($\frac{1}{50}$ inch).

Locality.—West coast of Patagonia (station 2784), 194 fathoms.

TRUNCATULINA PYGMÆA Hantken.

(Plate 77, fig. 6.)

Very small, slightly convex superiorily, quite convex inferiorily, and depressed at the center; rounded near the margin, but with a rather sharp edge; sutures sometimes thickened with clear shell substance. Diameter, about 0.36 mm. ($\frac{1}{70}$ inch).

Locality.—Gulf of Mexico (station 2460), 169 fathoms.

TRUNCATULINA ROSEA d'Orbigny.

(Plate 78, fig. 2.)

Superior face short conical, with rounded apex; inferior face flat or slightly convex; sutural lines very indistinct; color pink to bright rose color. Section shows chambers of the last convolution, and the thick deposit of pink shell substance about the center of the coil. Diameter, about 0.5 mm. ($\frac{1}{50}$ inch).

Locality.—Not recorded.

TRUNCATULINA PRÆCINCTA Karrer.

(Plate 78, fig. 1.)

Comparatively large, thick, biconvex, convexity greatest on the inferior side; margin obtuse; sutures raised by a thick deposit of clear shell substance, especially on the lower side and near the umbilicus. Diameter, about 1.5 mm. ($\frac{1}{17}$ inch).

Localities.—Gulf of Mexico (stations 2389, 2400), 169 and 196 fathoms.

aspects, give a fringed appearance to the margins of the segments. Aperture at the end of a short, oval, tubular neck, with a broad, everted edge. Diameter, about 0.5 mm. ($\frac{1}{50}$ inch).

Locality.—Gulf of Mexico (station 2352), 463 fathoms.

Genus ANOMALINA.

Characters similar to those of *Truncatulina*, except that the two faces are more nearly alike, the general contour being biconcave or subnautiloid, and the whole more or less evolute.

ANOMALINA AMMONOIDES Reuss.

(Plate 78, fig. 4.)

Symmetrical, about equally convex on the two faces, a little depressed at the umbilici, margin rounded; segments numerous, in three or four convolutions; sutures thickened with clear shell substance, sometimes a little raised; walls rather coarsely perforate; aperture in the middle line at the end of the last segment. Section has laid open every chamber of all the convolutions. Diameter, about 0.8 mm. ($\frac{1}{32}$ inch).

Locality.—Collected in large numbers off the west coast of Cuba (station 2352), 463 fathoms.

ANOMALINA GROSSERUGOSA Gümbel.

(Plate 78, fig. 5.)

Less symmetrical than *A. ammonoides*, superior face more compressed, segments fewer, and only those of the final convolution, about seven in number, visible on the inferior face. Diameter, about 1 mm. ($\frac{1}{25}$ inch).

Localities.—Gulf of Mexico and coast of Brazil (stations 2394, 2760), 420 and 1,019 fathoms.

ANOMALINA ARIMINENSIS d'Orbigny.

(Plate 79, fig. 1.)

Very much compressed, thin, margin square, with rounded angles; some of the earlier segments visible on the inferior face; sutures thick and sometimes prominent; walls transparent, distinctly showing outlines of segments and convolutions. Diameter, about 0.6 mm. ($\frac{1}{40}$ inch).

Locality.—Caribbean Sea (stations 2150, 2355), 382 and 399 fathoms.

ANOMALINA CORONATA Parker and Jones.

(Plate 79, fig. 2.)

Irregularly biconvex, the under side less convex than the upper, depressed at the center on both sides, often more or less distorted, the segments of the last convolution rapidly increasing in breadth, forming an irregular ridge around the border of each face; walls very coarsely porous. Diameter, about 0.25 mm. ($\frac{1}{100}$ inch).

Locality.—Off coast of Georgia (station 2416), 276 fathoms.

ANOMALINA POLYMORPHA Costa.

(Plate 79, fig. 3.)

Strongly resembles *A. coronata*, but is characterized by the presence of one or several short, stout spinous outgrowths, usually from the periphery of the shell. If but one spine is present, that is generally a prolongation of the final segment.

Locality.—Collected at the same station as *A. coronata.*

Genus RUPERTIA.

Test columnar, growing attached by a slightly-spreading base; segments numerous, spirally arranged; aperture at the inner margin of the final segment.

RUPERTIA STABILIS Wallich.

(Plate 79, fig. 4.)

Irregularly flask-shaped, having a moderately-inflated body, a short, thick neck, and an expanded lip. The lip is formed by the spreading base by which the shell adheres to some other body. The neck is formed by about two superimposed convolutions; the body by the inflated segments of the succeeding convolutions; walls thick and coarsely perforated; aperture at the inner edge of the final segment. Length, 1.5 mm. ($\frac{1}{16}$ inch), more or less.

Localities.—North Atlantic and Gulf of Mexico (stations 2530, 2383), 956 and 1,181 fathoms.

Subfamily TINOPORINÆ.

Test consisting of irregularly-heaped chambers, with a more or less distinctly spiral primordial portion.

Genus GYPSINA.

Test free or attached, spheroidal or spreading; structure acervuline, radiating, or laminated; chambers rounded or polyhedral, coarsely perforated.

GYPSINA INHÆRENS Schultze.

(Plate 79, fig. 6.)

Adherent; contour discoidal, more or less distorted according to the form of the surface to which it was adherent; composed of numerous subglobular segments irregularly heaped together, except at the very

Family X. NUMMULINIDÆ.

Test calcareous and finely tubulated; typically free, polythalamous and symmetrically spiral. The higher modifications all possessing a supplemental skeleton and a canal system of greater or less complexity.

Subfamily POLYSTOMELLINÆ.

Test bilaterally symmetrical; nautiloid, lower forms without supplemental skeleton or interseptal canals; higher types with canals opening at regular intervals along the external septal depressions.

Genus NONIONINA.

Supplemental skeleton absent or rudimentary; no external septal pores or bridges; aperture a simple curved slit.

NONIONINA BOUEANA d'Orbigny.

(Plate 79, fig. 5.)

Oval, compressed, bilaterally symmetrical; composed of numerous long, narrow, curved segments coiled in a close flat spiral, the last convolution completely inclosing the others; outline smooth; sutures flush; surface granular about the umbilici, which are depressed; no interseptal pores. Diameter, about 0.6 mm. ($\frac{1}{40}$ inch).

Locality.—Gulf of Tokyo, 9 fathoms.

NONIONINA SCAPHA Fichtel and Moll.

(Plate 80, fig. 1.)

Oval, compressed, symmetrical, smooth, not granular about the umbilici; segments comparatively few, increasing rapidly in size; face of the terminal segment broad and round. Diameter, about 0.4 mm. ($\frac{1}{60}$ inch).

Localities.—Panama Bay, coast of Yucatan, and Gulf of Tokyo (stations 2805, 2358), 9 to 222 fathoms.

Genus POLYSTOMELLA.

Supplemental skeleton, septal bridges, and canal system more or less fully developed; canals opening externally at the umbilicus and by a single or double row of pores along the sutures. Aperture a V-shaped line of perforations at the base of the septal face.

POLYSTOMELLA STRIATOPUNCTATA Fichtel and Moll.

(Plate 80, fig. 2.)

Discoidal, bilaterally symmetrical; final convolution incloses all the others; margin rounded; walls finely perforated; septal bridges distinct; a single row of pores along the sutures. Diameter, about 0.6 mm. ($\frac{1}{40}$ inch).

Localities.—Coast of Yucatan, North Atlantic (stations 2358, 2530, 2614), 10 to 956 fathoms.

POLYSTOMELLA CRISPA Linnæus.

(Plate 80, fig. 3.)

Lenticular, strongly biconvex, peripheral margin angular; septal pores in a single row, large, and closely set; umbilici filled with clear shell-substance more or less porous. Diameter, about 0.7 mm. ($\frac{1}{35}$ inch)

Locality.—Not recorded.

Subfamily NUMMULITINÆ.

Test lenticular or complanate: lower forms with thickened and finely tubulated shell-wall, but no intermediate skeleton; higher forms with interseptal skeleton and complex canal system.

Genus AMPHISTEGINA.

Test spiral, lenticular, inequilateral; chambers equitant, the alar prolongations on one side simple, on the other divided by deep constrictions so as to form supplementary lobes. Shell-wall thickened near the umbilicus and finely tubulated, but presenting no true canal system.

AMPHISTEGINA LESSONII d'Orbigny.

(Plate 80, fig. 4.)

Lenticular, somewhat unequally convex on the two sides; margin angular; surface smooth: segments numerous, narrow, bent, simple on the upper side, but constricted on the inferior side, and sharply bent backward; aperture on the under side of the last segment. Diameter, about 1.5 mm. ($\frac{1}{16}$ inch).

Localities.—North Atlantic, coast of Yucatan, Gulf of Mexico (stations 2415, 2629, 2641, 2363, 2370), 9 to 1,169 fathoms.

List of stations quoted, location, and depth of water.

tation.	Latitude.		Longitude.		Depth.	Locality.
	°	′	°	′	*Fathoms.*	
2040	38	35	68	16	2,226	Off Nantucket Shoals.
2041	39	22	68	25	1,608	Do.
2115	35	49	74	34	843	Off Cape Hatteras.
2117	15	24	63	31	683	Near Aves Island.
2144	9	49	79	31	896	Near Aspinwall.
2150	13	34	81	21	382	Near Old Providence Island.
2171	37	59	73	48	444	Off Maryland.
2204	39	30	71	44	728	South of Block Island.
2212	39	59	70	30	428	South of Marthas Vineyard.
2221	39	05	70	44	1,525	Do.
2225	36	05	69	51	2,512	Off North Carolina.
2228	37	25	73	06	1,582	Off Maryland.
2234	39	09	72	03	810	South of Long Island.
2242	40	15	70	27	58	South of Marthas Vineyard.
2243	40	10	70	26	63	Do.
2251	40	22	69	51	43	Off Nantucket Shoals.
2252	40	28	69	51	38	Do.
2263	37	08	74	33	430	Off Chesapeake Bay.
2264	37	07	74	34	167	Do.
2269	35	22	75	25	7	Off Cape Hatteras.
2312	32	54	77	53	88	Off South Carolina.
2313	32	53	77	53	99	Do.
2315	24	26	81	48	37	Off Key West, Florida.
2335	23	10	82	20	204	Off Habana, Cuba.
2338	23	10	82	20	189	Do.
2343	23	11	82	19	279	Do.
2352	22	35	84	23	463	Off west coast of Cuba.
2353	20	59	86	23	167	Arrowsmith Bank, Yucatan
2354	20	59	86	23	130	Do.
2355	20	56	86	27	399	Do.
2358	20	19	87	03	222	Off Cozumel Island, Yucatan.
2363	22	07	87	06	21	Off Cape Catoche, Yucatan.
2370	29	18	85	32	25	Between Delta of Mississippi River and Cedar Keys, Florida.
2372	29	15	85	29	27	Do.
2374	29	11	85	29	26	Do.
2377	29	07	88	08	210	Do.
2378	29	14	88	09	68	Do.
2379	28	00	87	42	1,467	Do.
2380	28	02	87	43	1,430	Do.
2382	28	19	88	01	1,255	Do.
2383	28	32	88	06	1,181	Do.
2385	28	51	88	18	730	Do.
2392	28	47	87	27	724	Do.
2394	28	38	87	02	420	Do.
2396	28	36	86	50	347	Do.
2398	28	45	86	26	227	Do.
2399	28	44	86	18	190	Do.
2400	28	41	86	07	169	Do.
2415	30	44	79	26	440	Off Georgia.

List of stations quoted, location, and depth of water—Continued.

Station.	Latitude.	Longitude.	Depth.	Locality.
	° '	° '	*Fathoms.*	
2416	31 26	79 07	276	Off Georgia.
2420	37 03	74 31	104	Off Chesapeake Bay.
2530	40 53	66 24	956	Southeast of Georges Bank.
2547	39 54	70 20	890	South of Marthas Vineyard.
2550	39 44	70 30	1,081	Do.
2565	38 19	69 02	2,020	About 220 miles southeast of Marthas Vineyard.
2568	39 15	64 09	1,781	About 300 miles southeast of Marthas Vineyard.
2569	39 26	68 03	1,782	Do.
2570	39 54	67 05	1,813	Southeast of Georges Bank.
2571	40 09	67 09	1,356	Do.
2576	41 13	68 15	18	Georges Bank.
2577	41 17	68 21	32	Do.
2584	39 05	72 23	341	South of Block Island.
2594	39 02	72 40	328	Do.
2614	34 09	75 02	168	Off Cape Lookout.
2616	33 42	77 31	17	Off Cape Fear.
2623	33 38	77 36	15	Do.
2627	32 21	77 07	437	Off Cape Romain.
2629	32 48	75 10	1,169	Mouth of Exuma Sound.
2641	25 11	80 10	60	Off Carysfort Light.
2650	23 34	76 34	360	Southeast of Andros Island (Bahamas).
2651	24 02	77 12	97	Do.
2654	27 57	77 27	660	Off Little Bahama Bank.
2655	27 22	78 07	338	Do.
2660	28 40	78 46	504	Off Cape Canaveral.
2662	29 24	79 43	434	Off St. Augustine.
2663	29 39	79 49	421	Do.
....	....	....	....	
2677	32 39	76 50	478	Off Cape Fear.
2679	32 40	76 40	782	Do.
2684	39 35	70 54	1,106	South of Marthas Vineyard.
2716	38 29	70 57	1,631	Do.
2723	36 47	73 09	1,685	Off Chesapeake Bay.
2731	36 45	74 28	781	Do.
2750	18 30	63	496	Off Windward Islands, West Indies.
2751	16 54	63 12	687	Do.
2754	11 40	58 33	880	Off Santa Lucia, West Indies.
2760	S. 12 07	37 17	1,019	Off Bahia, Brazil.
2762	S. 23 08	41 34	59	Off Cape Frio, Brazil.
2763	S. 24 17	42 48	671	Do.
2784	S. 48 41	74 24	194	Between Wellington Island and Patagonia.
2805	S. 7 56	79 41	51	Panama Bay.
2842	N. 54 15	166 03	72	Off Head of Akutan Island, Alaska.
[illegible]	[illegible]	[illegible]	[illegible]	Off Cape St. James, Queen Charlotte Islands.

INDEX.

	Plate.	Figure.	Page.
Allomorphina			263
Alveolina			262
Alveolininae			262
Ammodiscus			260, 278
charoides	24	2	279
gordialis	24	1	279
incertus	23	2	278
tenuis	23	1	279
Amphicoryne			262
Amphimorphina			262
Amphistegina			264, 338
lessonii	80	4	338
Anomalina			263, 335
ammonoides	78	4	335
ariminensis	79	1	335
coronata	79	2	335
grosserugosa	78	5	335
polymorpha	79	3	336
Aphrosina			264
Archædiscus			264
Articulina			261, 301
sagra	47	1	301
Aschemonella			250
Assilina			264
Astrorhiza			259, 265
angulosa	3	1	265
arenaria	3	2	265
crassatina	2		265
granulosa	1		265
Astrorhizidæ			258, 259, 264
Astrorhizinæ			265
Bathysiphon			259, 267
rufus	7		267
Bdelloidina			260
Bifarina			261
Bigenerina			261, 286
capreolus	32	3	286
nodosaria	31	4	286
pennatula	32	2	287
robusta	32	1	286
Biloculina			261
bulloides	38	5	293
comata	39	3	294
dehiscens	40	3	295
depressa	40	1	294
var. serrata	40	2	294
elongata	39	4	294
irregularis	41	3	295
lævis	41	1	295
ringens	39	2	294
sphæra	41	2	296
tubulosa	39	1	293

	Plate.	Figure.	Page.
Bolivina			261, 291
ænariensis	37	8	292
porrecta	38	2	292
punctata	38	1	292
Botellina			259
Bradyina			260
Bulimina			261, 290
aculeata	37	4	291
affinis	37	2	290
elegans	36	3	290
inflata	37	5	291
pupoides	37	3	290
pyrula	36	4, 5	290
var. spinescens	37	1	290
Buliminin␣æ			261, 289
Calcarina			263
Candeina			263, 325
nitida	71	3	325
Carpenteria			263
Carterina			260
Cassidulina			261, 292
crassa	38	3	292
subglobosa	38	4	293
Cassidulininæ			261, 292
Chilostomella			263
Chilostomellidæ			258, 263
Chrysalidina			261
Clavulina			261, 288
angularis	36	2	289
communis	34	3	288
eocæna	35	1	289
parisiensis	35	2	289
var. humilis	36	1	289
Cornuspira			262, 303
foliacea	48	1	303
involvens	48	3	303
Cristellaria			262, 314
aculeata	66	3	318
acutauricularis	63	5	316
articulata	64	2	317
calcar	66	1	318
compressa	62	1	315
crepidula	63	2	316
cultrata	65	2	318
echinata	66	2	318
gibba	64	1	317
italica	63	6	316
latifrons	63	3	316
limbata	67	1	318
obtusata var. subulata	61	3	315

	Plate.	Figure.	Page.
ionina			259, 266
hispida	6	2	267
pisum	6	1	266
olina			260
mmina			260, 282
cancellata	27	3	282
"	28	1	
pusilla	28	2	282
clypeinae			264
clypeus			263, 326
alopora			263, 326
poeyi	72	1	326
rophrya			259
alinopsis			262
oropodon			258
rphina			262
rbina			263, 327
bertholoti	72	4	327
biconcava	72	5	327
globularis	72	2	327
rosacea	72	3	327
nbergia			261
soidina			263
thyra			260
thyrinæ			260
laria			261
ellina			262
minifera			252, 258
dicularia			262, 313
alata	59	1	313
inæqualis	59	2	313
lina			264
lininæ			264
ryina			261, 287
baccata	32	5	287
filiformis	33	2	287
pupoides	32	4	287
rugosa	33	3	288
scabra	34	1	288
siphonella	34	2	288
subrotundata	33	1	287
igerina			263, 321
æquilateralis	70	3	323
bulloides	69	2	321
conglobata	69	6	322
digitata	70	2	323
dubia	69	4	322
inflata	69	3	322
rubra	69	5	322
sacculifera	70	1	322
igerinidæ			258, 263, 321
nia			258
nidæ			258
ina			264, 336
inhærens	79	6	336
physema			259
ophragmium			259, 275

	Plate.	Figure.	Page.
Haplophragmium agglutinans	19	2	275
calcareum	19	1	275
canariense	20	3	277
cassis	19	4	275
emaciatum	19	5	276
foliaceum	19	6	276
globigeriniforme	21	1	277
latidorsatum	20	1	276
scitulum	20	2	276
tenuimargo	19	3	275
Haplostiche			260, 277
soldanii	21	3	277
Hastigerina			263, 323
pelagica	70	4	324
Haueriua			261
Haueriuinæ			262, 261
Heterostegina			264
Hipporepina			260
Hormosina			260, 280
carpenteri	25	1	280
globulifera	24	4	280
ovicula	25	2	280
Hyperammina			259, 269
elongata	10	2	270
friabilis	10	1	269
ramosa	11	1	270
vagans	11	2	270
Involutina			260
Jaculella			259, 269
acuta	9	4	269
Keramosphæra			262
Keramosphærinæ			262
Lagena			262, 305
castanea	54	3	307
castrensis	54	5	308
distoma	53	5	306
elongata	53	1	306
globosa	53	4	306
gracillima	53	3	306
hispida	53	8	307
levis	53	6	306
longispina	53	2	306
marginata	54	2	307
orbignyana	54	4	308
staphyllearia	54	1	307
sulcata	53	7	307
Lagenidæ			258, 262, 305
Lieberkühnia			258
Lingulina			262, 312
carinata	58	3	312

	Plate.	Figure.	Page.
..sinæ			260, 282
..inulina			262, 313
ensis	59	3	314
glabra	60	1	313
ıpella			259, 270
elongata	10	2	270
ogromia			258
lidæ			258, 261, 293
lina			261, 297
agglutinans	47	2	301
angularis	46	1	300
auberiana	43	6	298
bicornis	46	2	300
bucculenta	45	1	299
circularis	44	[illegible]	298
cuvierana	43	[illegible]	298
gracilis	43	[illegible]	297
insignis	45	2	299
labiosa	45	[illegible]	299
linnæana	46	[illegible]	300
oblonga	43	[illegible]	297
pulchella	46	[illegible]	301
reticulata	46	[illegible]	301
seminulum	43	[illegible]	297
separans	46	[illegible]	300
subrotunda	44	[illegible]	299
tricarinata	44	[illegible]	298
trigonula	44	[illegible]	298
undosa	45	[illegible]	300
valvularis	44	[illegible]	299
venusta	44	[illegible]	298
lininæ			261, 293
saria			262, 308
catenulata	58	[illegible]	312
comata	57	[illegible]	311
communis	56	2	310
consobrina var. emaciata	56	1	310
costulata	58	[illegible]	312
farcimen	55	[illegible]	309
filiformis	55	[illegible]	310
hispida	57	[illegible]	311
var. sublineata	56	4	311
lævigata	55	[illegible]	308
mucronata	57	[illegible]	311
obliqua	57	4	311
pyrula	55	4	309
radicula	55	1	309
rœmeri	56	5	310
rotundata	54	6	308
simplex	55	2	309
soluta	56	3	310
vertebralis	57	5	312
sarinæ			262, 308
sinella			260
›nina			264, 337
boueana	79	5	337

	Plate.	Figure.	Page.
Nonionina scapha	80	1	337
Nubecularia			261
Nubecularinæ			261
Nummulinidæ			258, 264, 337
Nummulites			264
Nummulitinæ			264, 338
Operculina			264
Ophthalmidium			261, 302
inconstans	47	3	302
Orbiculina			262, 304
adunca	50	1	304
Orbitoides			264
Orbitolites			262, 304
duplex	51	2, 3	305
marginalis	50	2	304
"	51	1	
tenuissima	52		305
Orbulina			263, 323
universa	69	1	323
Parkeria			260
Patellina			263
Pavonina			261
Pelosina			259, 265
variabilis	4	1	266
Peneroplidinæ			262, 303
Peneroplis			262, 303
pertusus	48	4	304
var. discoideus	49	1	304
Pilulina			259, 266
jeffreysii	5		266
Pilulininæ			259, 266
Placopsilina			260
Planispirina			262, 302
celata	47	5	303
sigmoidea	47	6	302
Planorbulina			263, 328
acervalis	72	7	328
mediterranensis	72	6	328
Pleurostomella			261
Polymorphina			262, 319
communis	67	6	319
compressa	67	3	319
elegantissima	67	4	319
oblonga	67	5	319
sororia var. fistulosa	67	2	319
Polymorphininæ			262, 319
Polyphragma			260
Polystomella			264, 337
crispa	80	3	338
striatopunctata	80		337

	Plate.	Figure.	Page.
Psammosphæra parva	9	1	268
Pullenia			263, 324
obliquilobulata	70	6	324
quinqueloba	70	5	324
Pulvinulina			263, 328
auricula	73	2	329
crassa	74	1	329
elegans	75	1	331
karsteni	74	5	330
menardii	73	3	329
var. fimbriata	73	4	329
micheliana	74	2	330
partschiana	75	3	331
pauperata	74	3	330
punctulata	73	1	328
repanda	72	8	328
tumida	73	5	329
umbonata	74	4	330
Ramulina			263, 321
globulifera	68	6	321
proteiformis	68	7	321
Ramulininæ			263, 321
Reophax			259, 272
adunca	18	5	274
bacillaris	18	3	274
bilocularis	17	2	273
cylindrica	18	6	274
dentaliniformis	18	2	274
difflugiformis	16	2	272
var. testacea	16	1	273
nodulosa	18	4	274
pilulifera	18	1	273
scorpiurus	16	3	273
Rhabdammina			259, 271
abyssorum	12	2	271
cornuta	14	2	271
discreta	13	....	271
linearis	14	1	271
Rhabdammininæ			259, 269
Rhabdogonium			262
Rhizammina			259, 272
algæformis	15	1	272
indivisa	15	2	272
Rhizopoda			258
Rimulina			262
Rotalia			263, 331
beccarii	75	2	331
orbicularis	75	5	331
papillosa	76	2	332
pulchella	76	3	332
schrœteriana	76	1	332
soldanii	75	4	332
Rotalidæ			258, 263, 325
Rotalinæ			263, 326
Rupertia			263, 336
stabilis	79	4	336

	Plate.	Figure.	Page.
Saccammina			259, 268
conscociata	9	3	269
sphærica	9	2	269
Saccammininæ			259, 267
Sagenella			259
Sagrina			262
Schwagerina			264
Shepheardella			258
Sorosphæra			259
Sphæroidina			263, 324
bulloides	71	1	325
dehiscens	71	2	325
Spirillina			263, 325
limbata	71	5	326
obconica	71	6	326
vivipara	71*	4	326
Spirillininæ			263, 325
Spiroloculina			261, 295
arenaria	43	1	297
excavata	41	5	296
limbata	42	3	296
nitida	41	4	296
planulata	42	4	297
robusta	42	1	296
series from biloculina	42	2	296
Spiroplecta			261
Squamulina			261
Stacheia			260
Storthosphæra			259, 266
albida	4	2	266
Syringammina			259
Technitella			259
Textularia			260, 283
agglutinans	29	4	284
barrettii	30	2	285
carinata	29	1	284
concava	28	5	283
conica	29	6	285
gramen	29	5	284
luculenta	29	3	284
quadrilatera	28	3	283
rugosa	29	2	284
transversaria	28	4	283
trochus	30	1	285
Textularidæ			258, 260, 283
Textularinæ			260, 283
Thalamopora			264
Thurammina			260, 278
[illegible]	22	2	278
favosa	27	2	278

	Plate.	Figure.	Page.
Trochammina coronata	26	3	281
lituiformis	26	1	281
pauciloculata	27	2	282
proteus	25	3	281
ringens	27	1	281
Trochammininæ			260, 277
Truncatulina			263, 332
akneriana	77	5	333
lobatula	76	4	333
præcincta	78	1	334
pygmæa	77	6	334
reticulata	78	3	334
robertsoniana	77	3	333
rosea	78	2	334
tenera	77	4	334
ungeriana	77	2	333
wuellerstorfi	77	1	333
Uvigerina			262, 320
angulosa	68	3	320
asperula	68	4	320
var. **ampullacea**	68	5	320
pygmæa	68	2	320
tenuistriata	68	1	320
Vaginulina			262, 314
legumen	60	2	314
linearis	61	1	314
spinigera	60	3	314
Valvulina			261, 286
conica	31	3	286
Verneuilina			261, 285
propinqua	31	2	285
pygmæa	31	1	285
Vertebralina			261, 301
insignis	47	4	302
Virgulina			261, 291
schreibersiana	37	6	291
subsquamosa	37	7	291
Webbina			260, 279
clavata	24	3	279

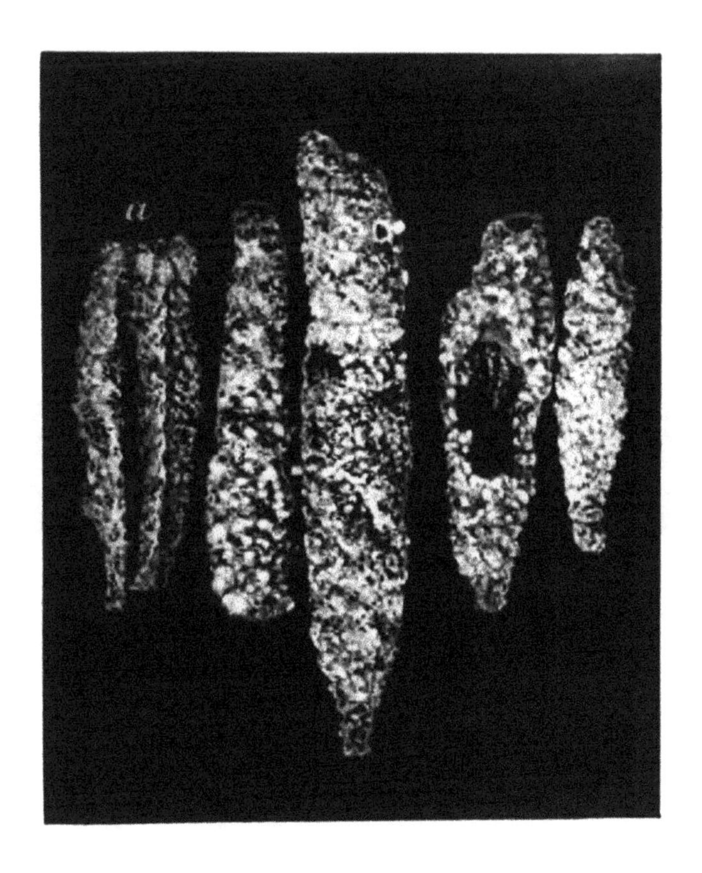

ASTRORHIZA GRANULOSA BRADY. SEE PAGE 265

a Longitudinal Section.

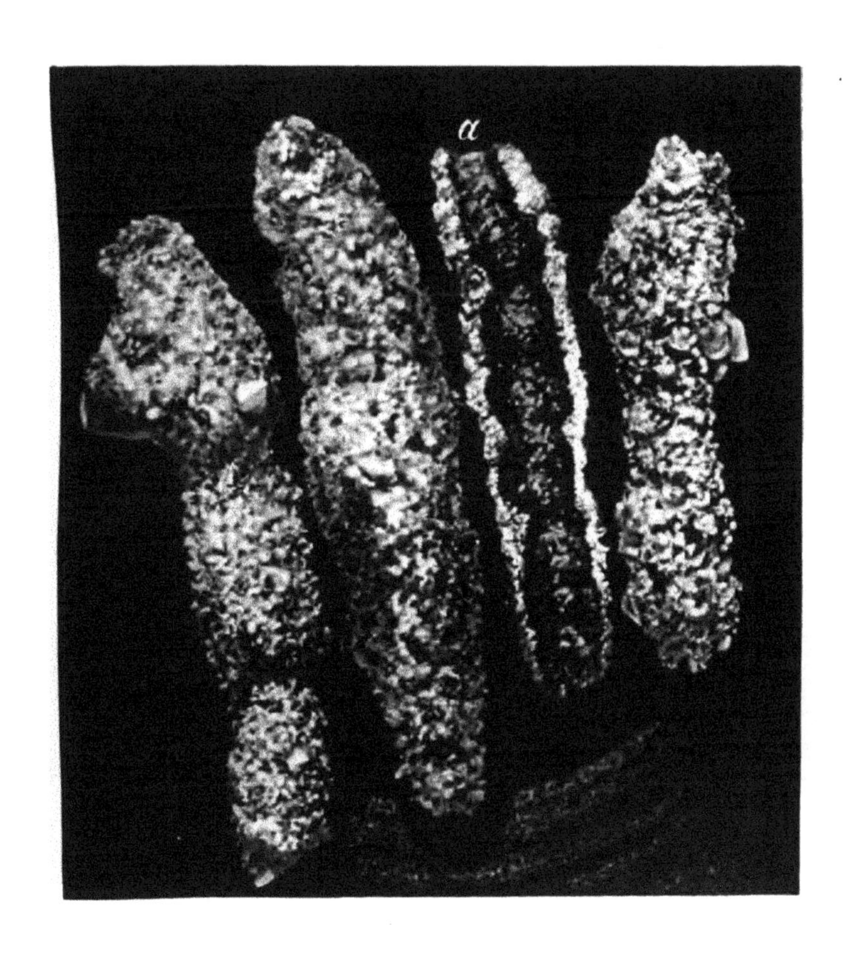

ASTRORHIZA CRASSATINA BRADY. SEE PAGE 265.
a. Longitudinal Section

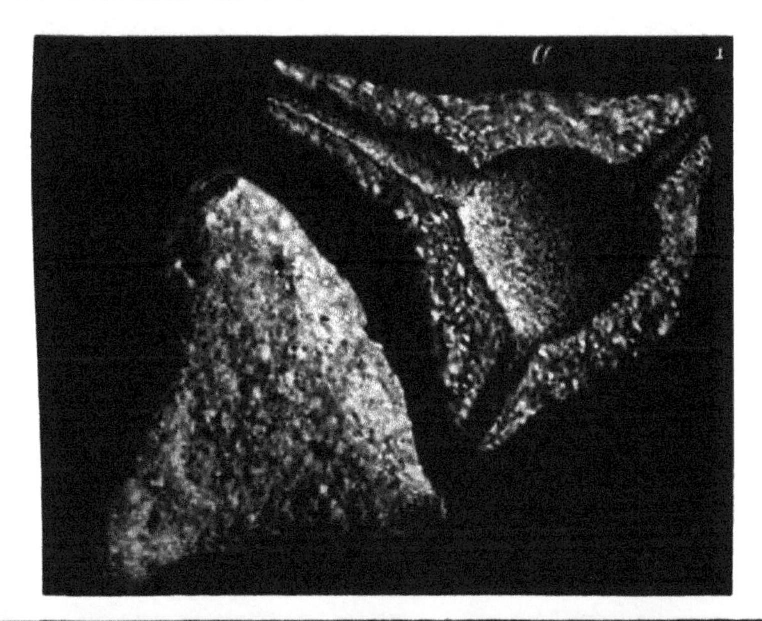

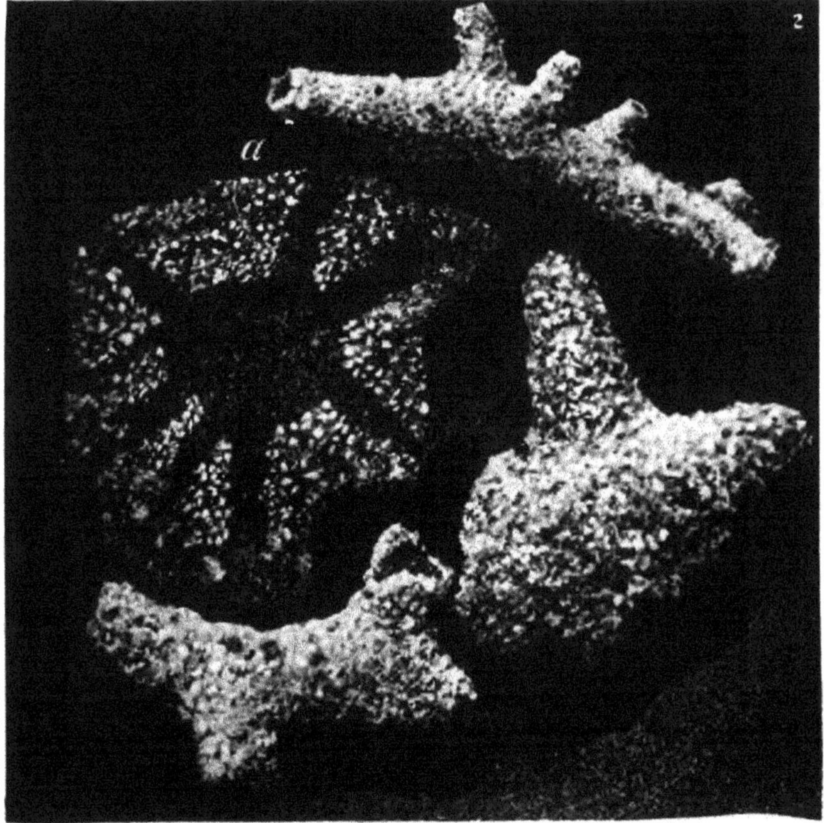

FIG. 1. ASTRORHIZA ANGULOSA BRADY. SEE PAGE 265.

a. Section.

FIG. 2. ASTRORHIZA ARENARIA NORMAN. SEE PAGE 265.

a. Section.

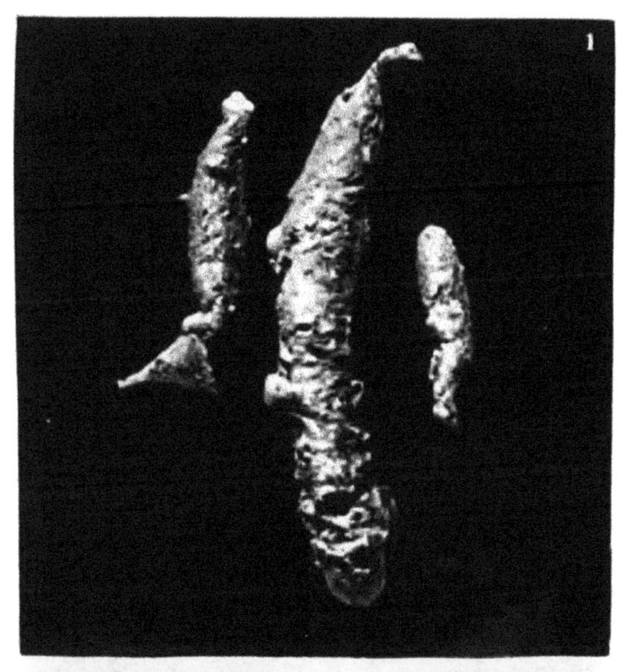

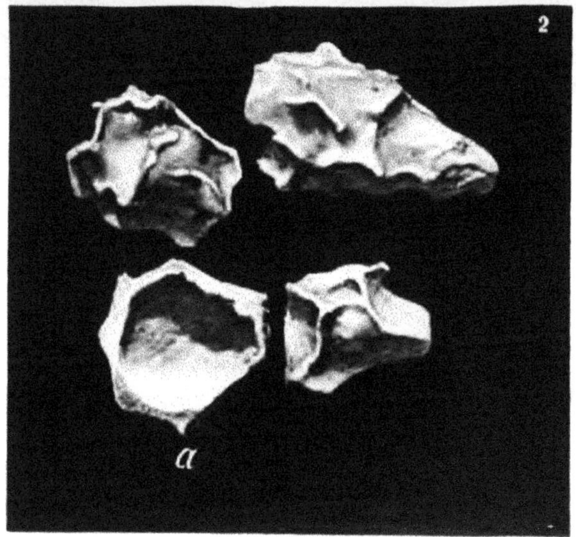

FIG. 1. PELOSINA VARIABILIS BRADY. SEE PAGE 266.
FIG. 2. STORTHOSPHAERA ALBIDA SCHULTZE. SEE PAGE 266.
a. Section.

PILULINA JEFFREYSII CARPENTER. SEE PAGE 266.

a. Section

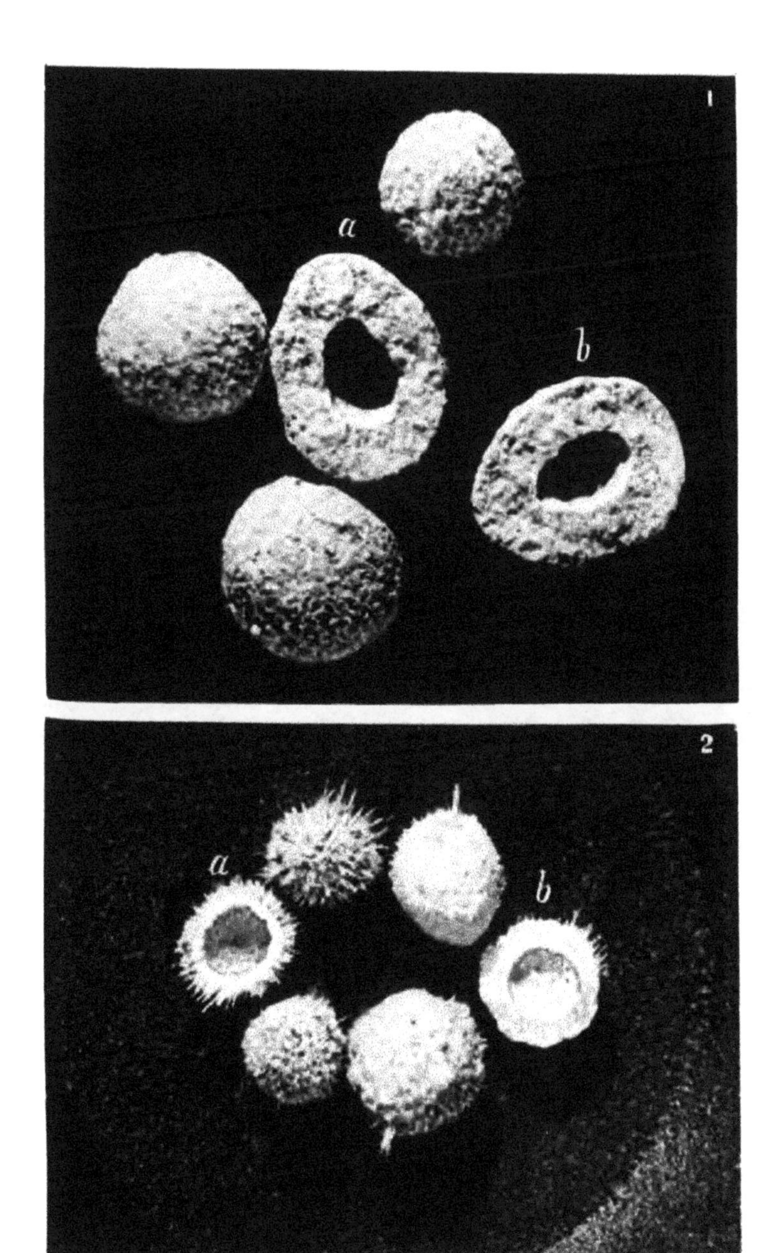

FIG. 1. CRITHIONINA PISUM GOËS. SEE PAGE 266.
a, b. Sections.

FIG. 2. CRITHIONINA PISUM GOËS, VAR. HISPIDUM, NEW. SEE PAGE 267.
a, b. Sections.

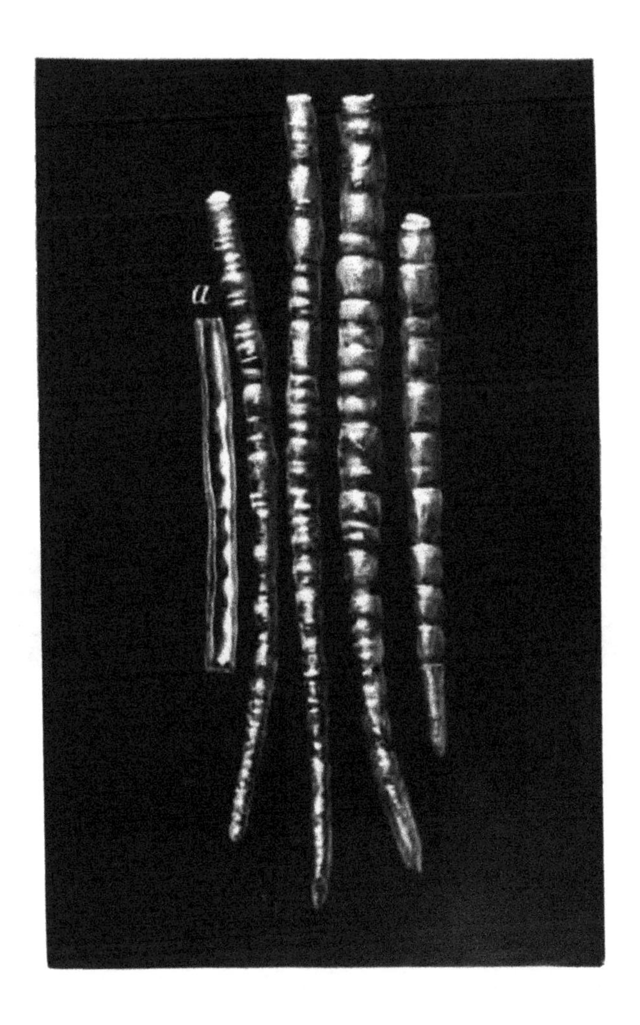

BATHYSIPHON RUFUM DE FOLIN. SEE PAGE 267.

a. Longitudinal Section.

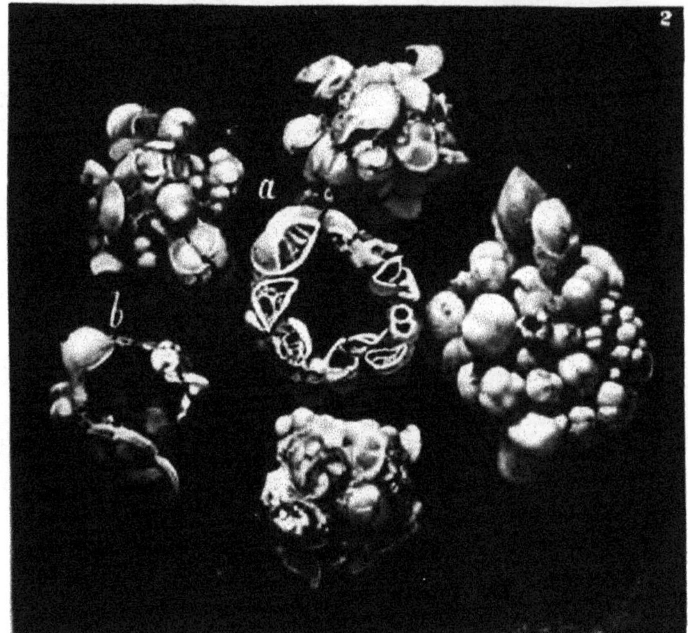

FIG. 1. PSAMMOSPHAERA FUSCA BRADY. SEE PAGE 268.

a, b. Sections.

FIG. 2. PSAMMOSPHAERA FUSCA BRADY, VAR. TESTACEA, NEW. SEE PAGE 268.

a. Artificial Section. b. Accidental Section.

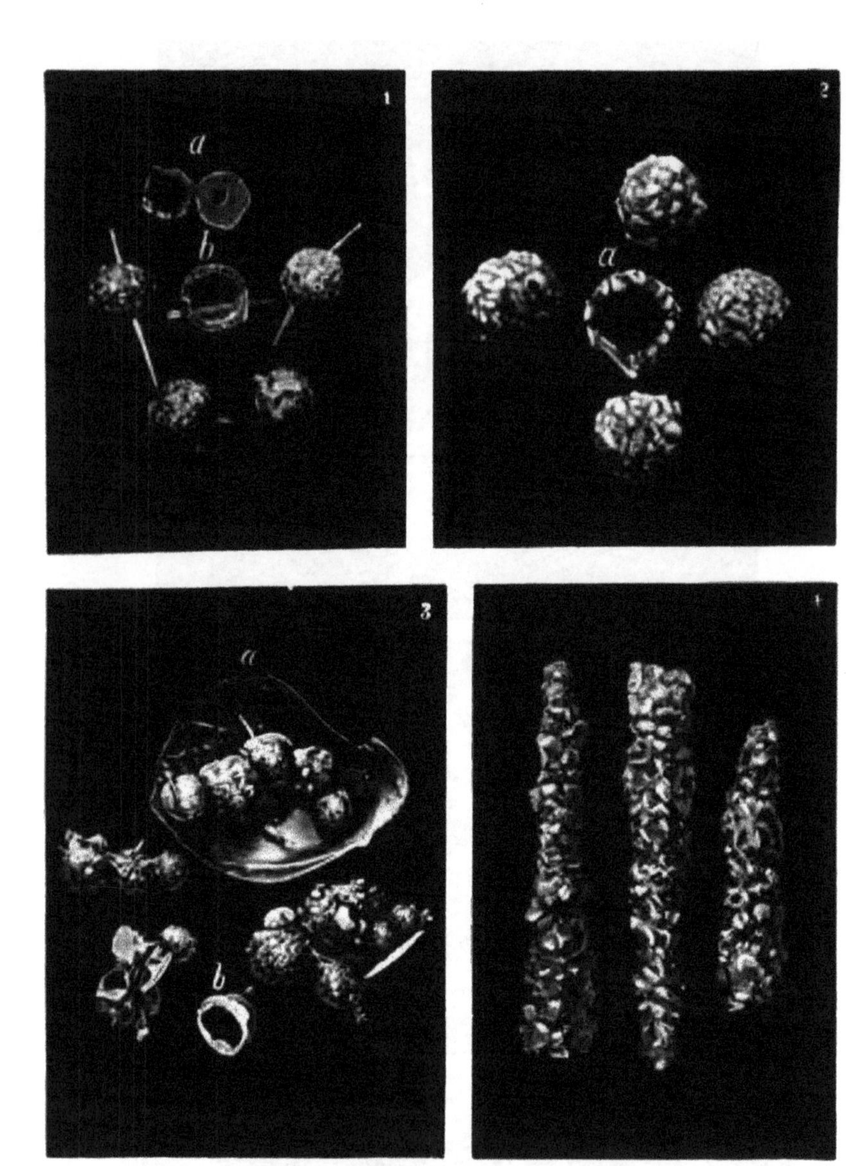

FIG. 1. PSAMMOSPHÆRA PARVA M. SARS. SEE PAGE 268.
a Adherent Specimen.

FIG. 2. SACCAMMINA SPHERICA M. SARS. SEE PAGE 269.
a. Section.

FIG. 3. SACCAMMINA CONSOCIATA NEW SPECIES. SEE PAGE 269.
a. Adherent to a fragment of shell. b. Detached Specimen

FIG. 4. JACULELLA ACUTA BRADY. SEE PAGE 269.

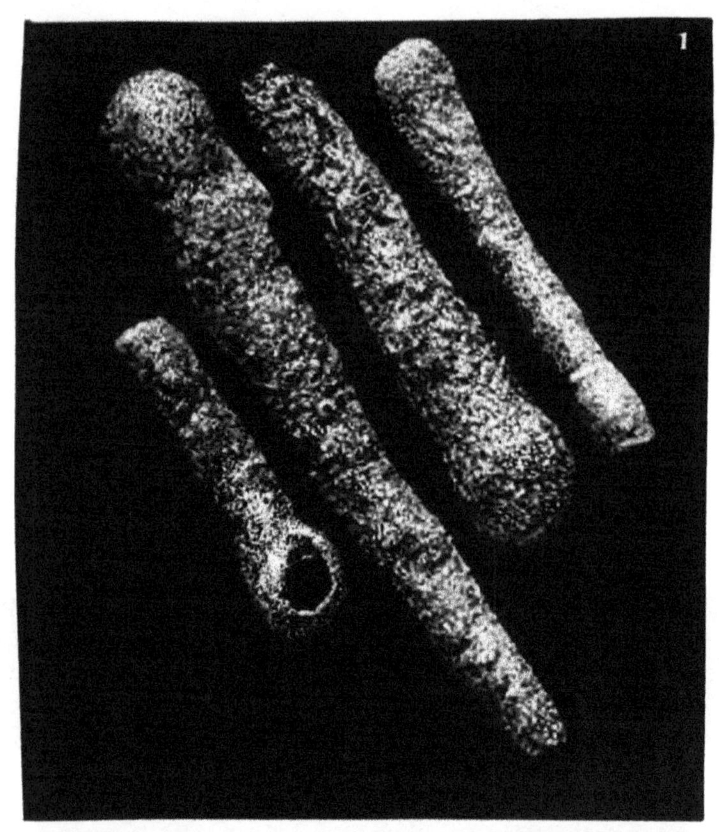

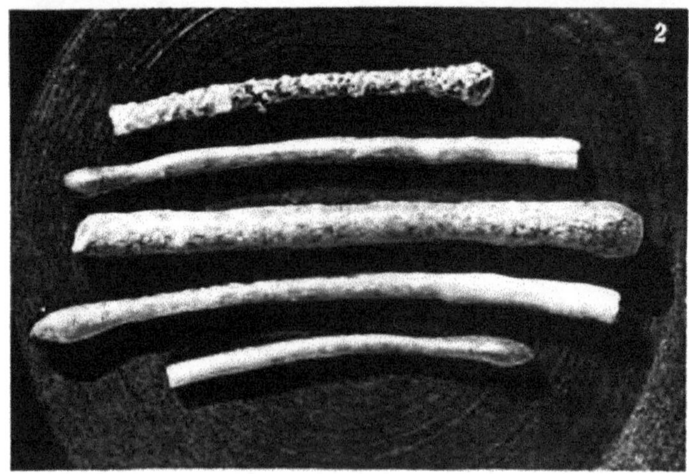

FIG. 1. HYPERAMMINA FRIABILIS BRADY. SEE PAGE 269.

FIG. 2. HYPERAMMINA ELONGATA BRADY. SEE PAGE 270.

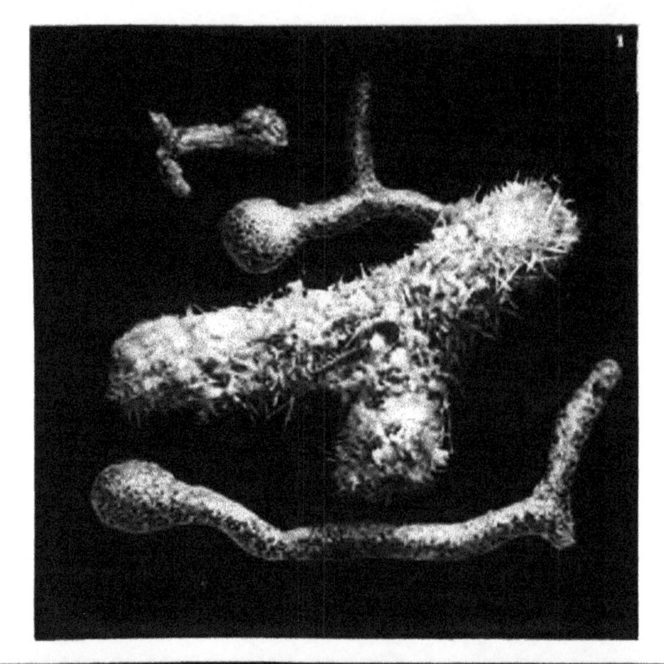

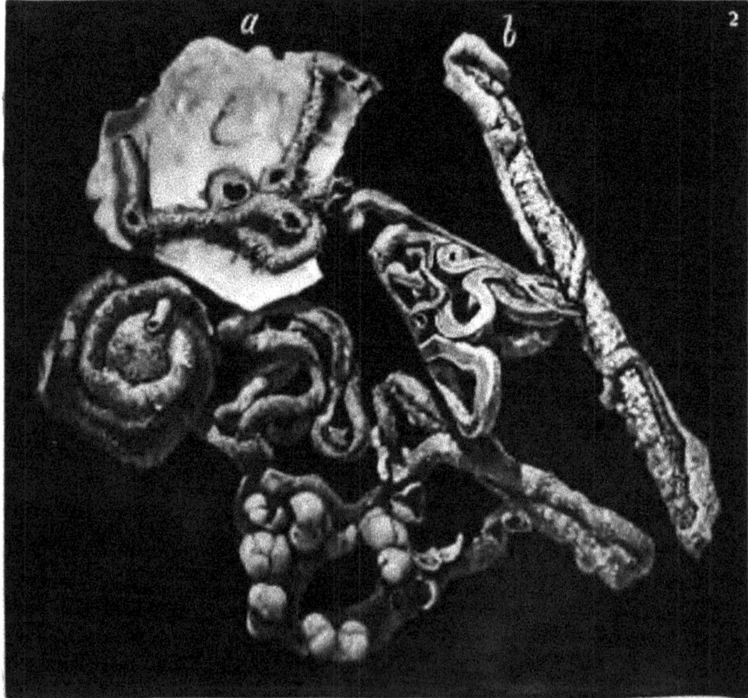

FIG. 1. HYPERAMMINA RAMOSA BRADY. SEE PAGE 270.

FIG. 2. HYPERAMMINA VAGANS BRADY. SEE PAGE 270.

a. Specimen attached to fragment of Shell of Mollusk.

b. Specimen coiled around fragment of Rhabdammina

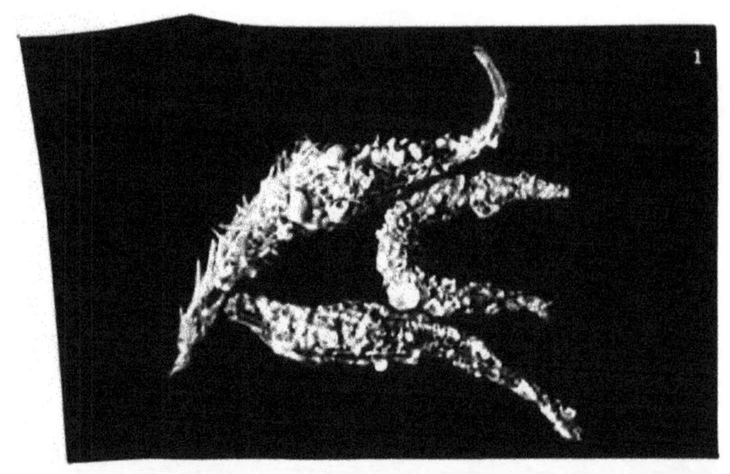

FIG. 1. MARSIPELLA ELONGATA NORMAN. SEE PAGE 270.

FIG. 2. RHABDAMMINA ABYSSORUM M. SARS. SEE PAGE 271.

a. Section.

RHABDAMMINA DISCRETA BRADY. SEE PAGE 271.

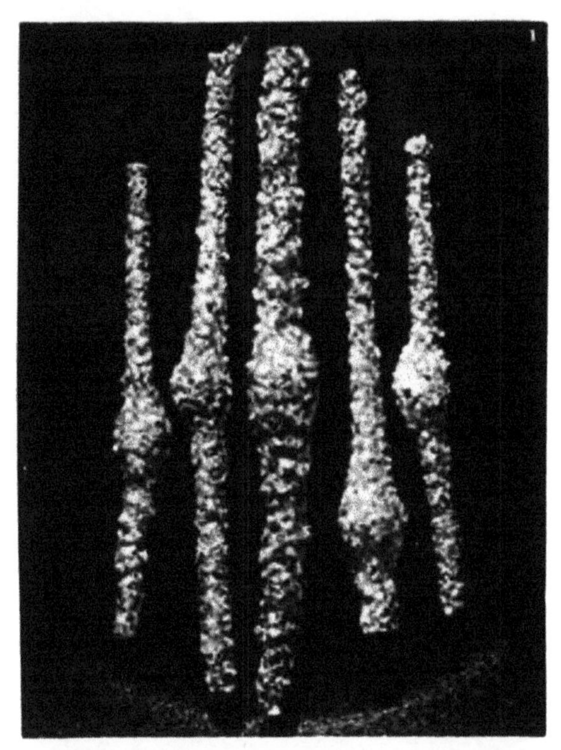

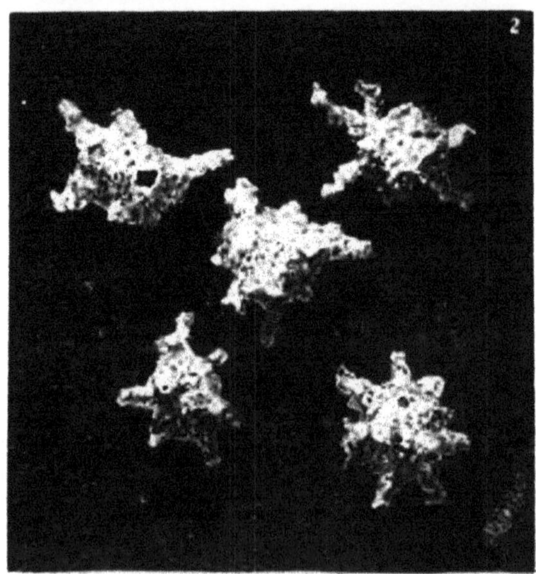

FIG. 1. RHABDAMMINA LINEARIS BRADY. SEE PAGE 271.
FIG. 2. RHABDAMMINA CORNUTA BRADY. SEE PAGE 271.

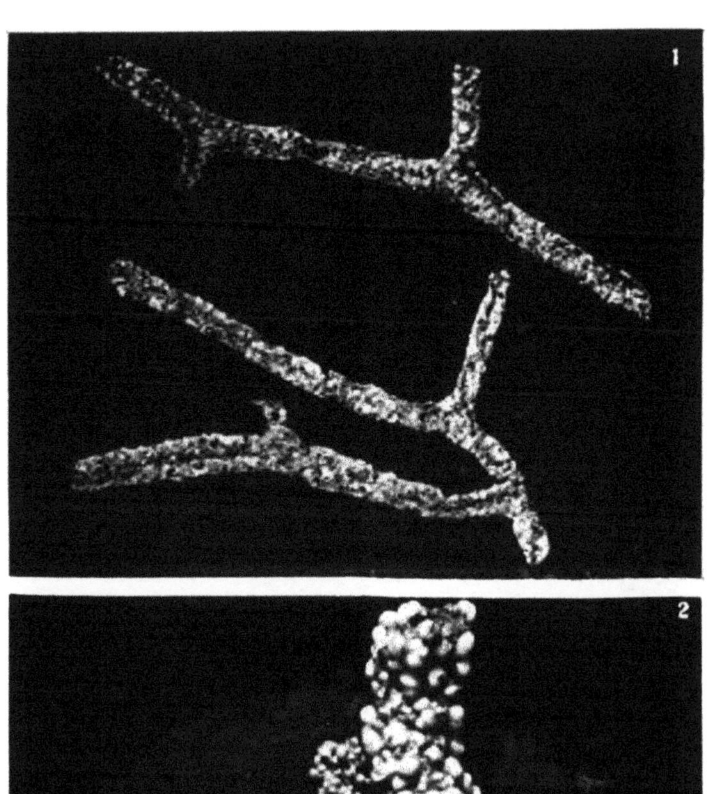

FIG. 1. RHIZAMMINA ALGÆFORMIS BRADY. SEE PAGE 272.
FIG 2. RHIZAMMINA INDIVISA BRADY. SEE PAGE 272.

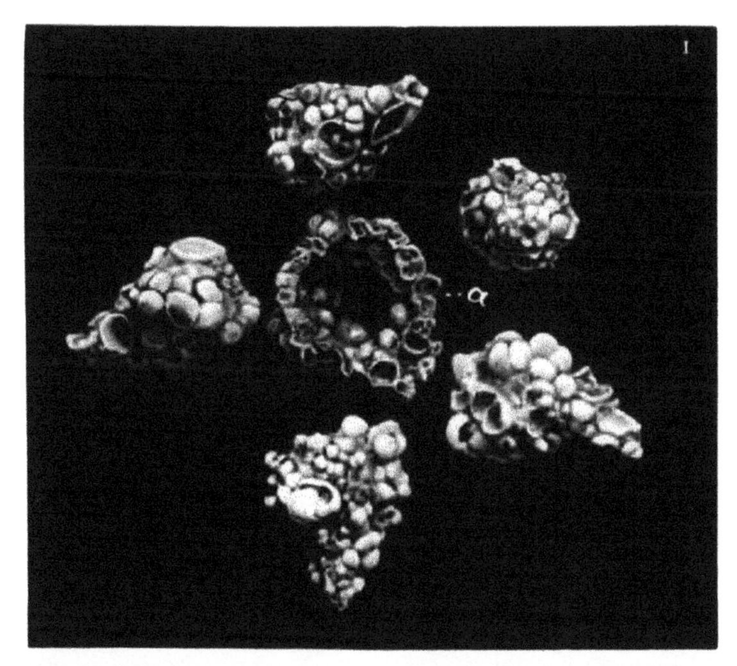

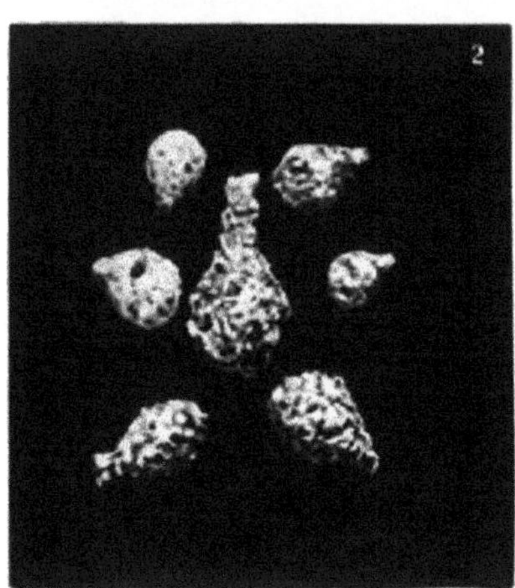

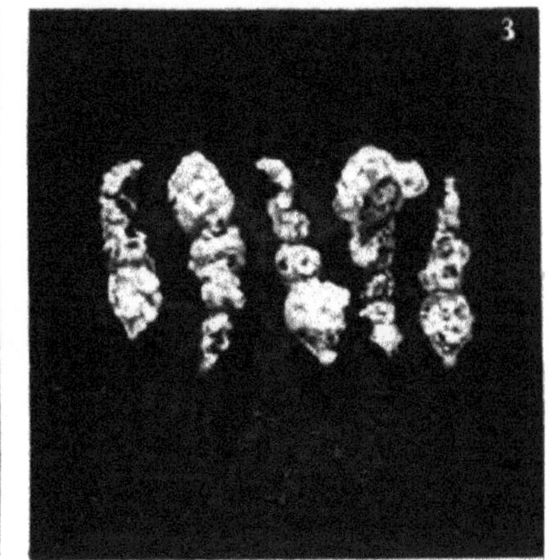

FIG. 1. REOPHAX DIFFLUGIFORMIS BRADY, VAR. TESTACEA, NEW. SEE PAGE 273.

a. Longitudinal Section.

FIG. 2. REOPHAX DIFFLUGIFORMIS BRADY. SEE PAGE 272.

FIG. 3. REOPHAX SCORPIURUS MONTFORT. SEE PAGE 273.

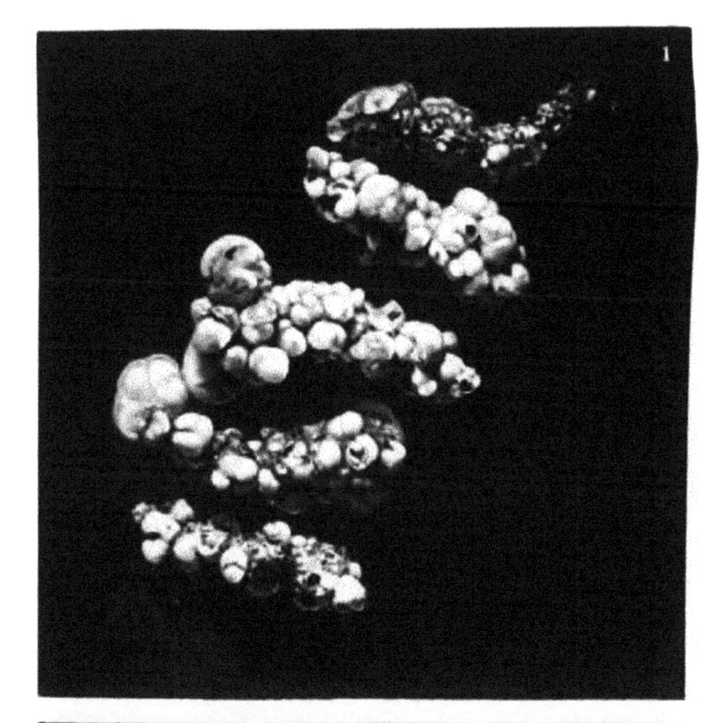

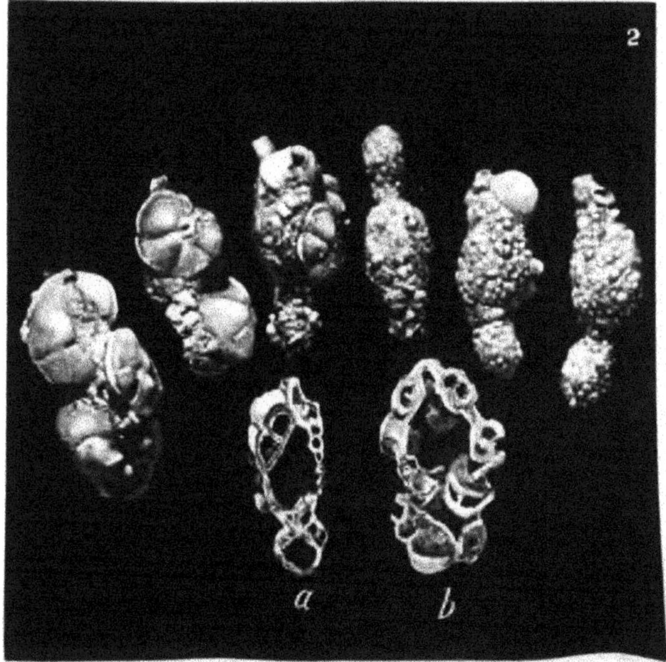

FIG. 1. REOPHAX SCORPIURUS MONTFORT. SEE PAGE 273.
FIG. 2. REOPHAX BILOCULARIS NEW SPECIES. SEE PAGE 273.
a, *b*. Longitudinal Sections

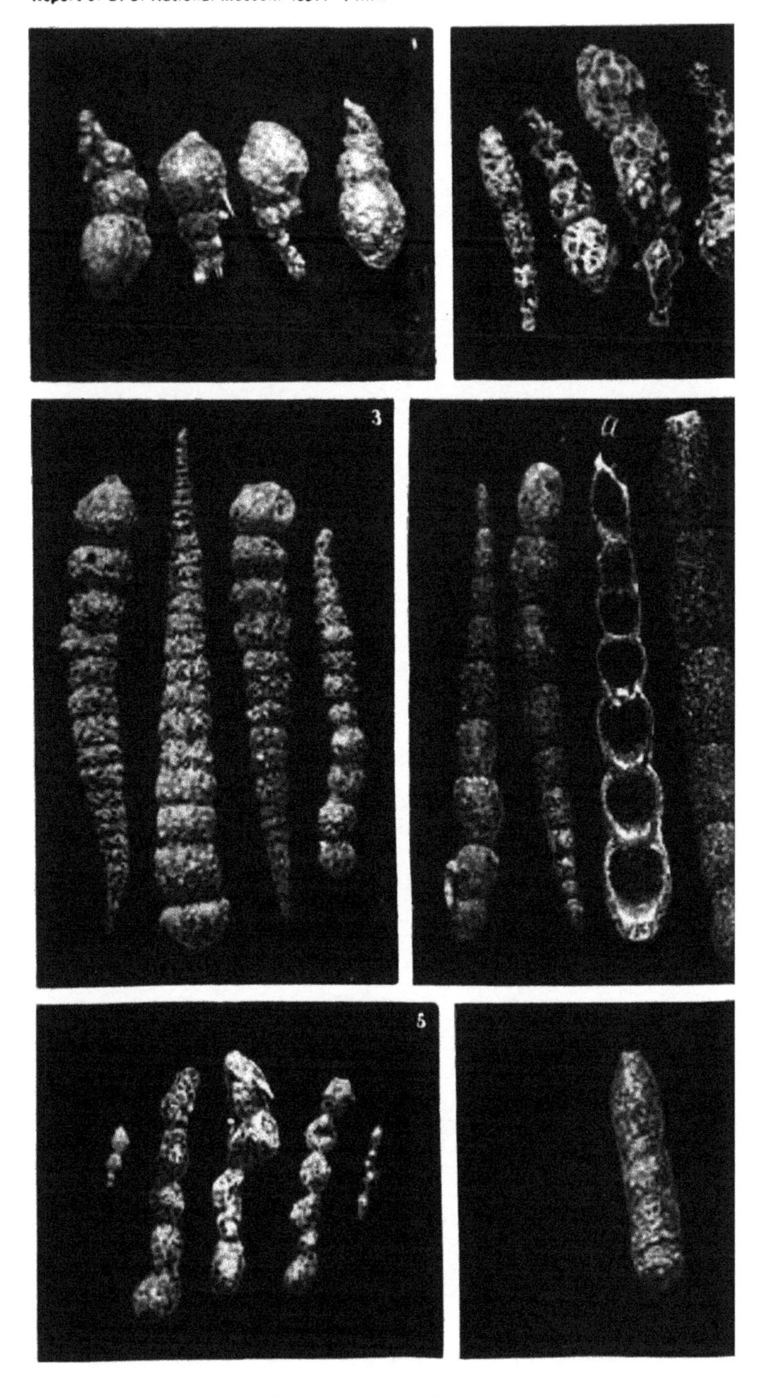

FIG. 1. REOPHAX PILULIFERA BRADY. SEE PAGE 273.
FIG. 2. REOPHAX DENTALINIFORMIS BRADY. SEE PAGE 274.
FIG. 3. REOPHAX BACILLARIS BRADY. SEE PAGE 274.
FIG. 4. REOPHAX NODULOSA BRADY. SEE PAGE 274.
a. Longitudinal Section
FIG. 5. REOPHAX ADUNCA BRADY. SEE PAGE 274.

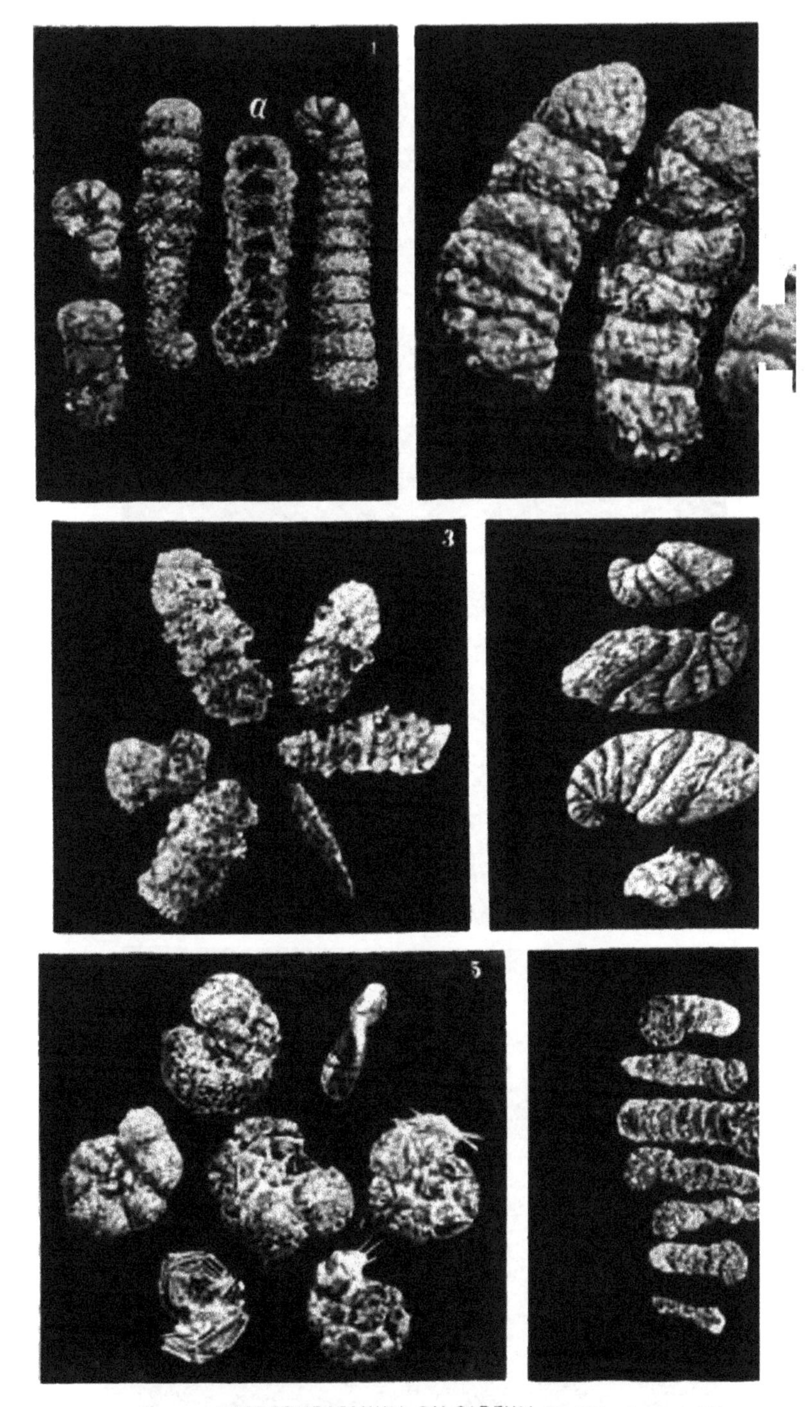

Fig. 1. HAPLOPHRAGMIUM CALCAREUM Brady. See Page 275.

a. Longitudinal Section.

Fig. 2. HAPLOPHRAGMIUM AGGLUTINANS Brady. See Page 275.

Fig. 3. HAPLOPHRAGMIUM TENUIMARGO Brady. See Page 275.

Fig. 4. HAPLOPHRAGMIUM CASSIS Parker. See Page 275.

Fig. 5. HAPLOPHRAGMIUM EMACIATUM Brady. See Page 276.

Fig. 6. *HAPLOPHRAGMIUM* FOLIACEUM Brady. See Page 276.

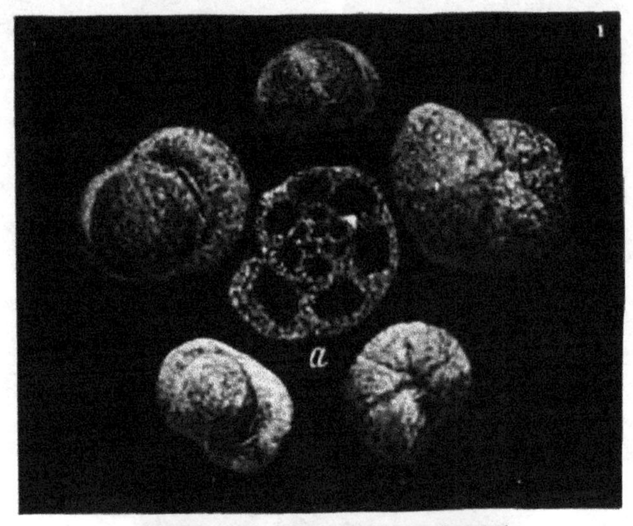

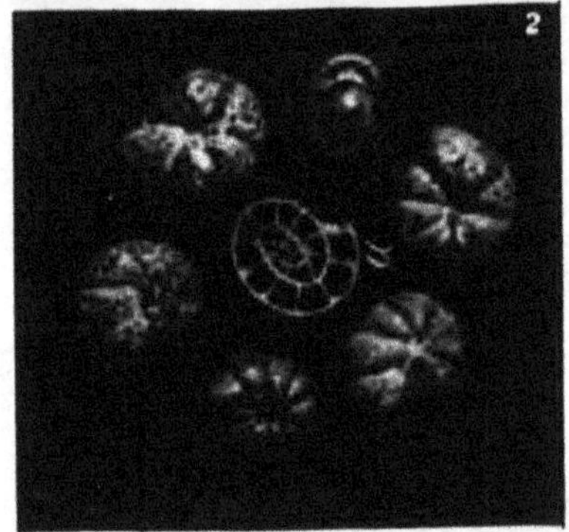

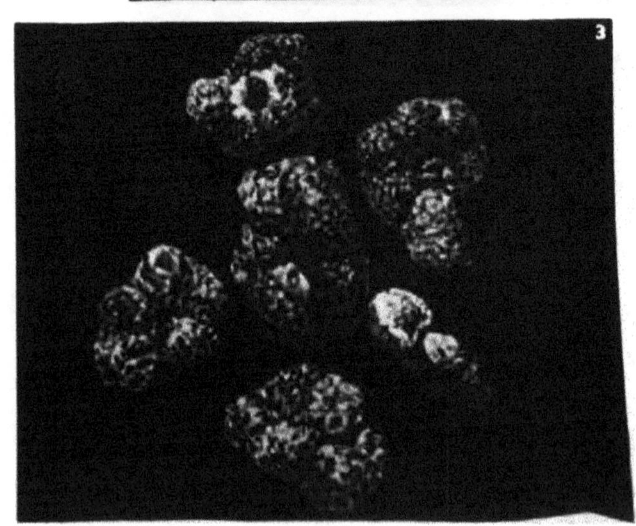

FIG. 1. HAPLOPHRAGRMIUM LATIDORSATUM BORNEMANN. SEE PAGE 276.

FIG. 2. HAPLOPHRAGRMIUM SCITULUM BRADY. SEE PAGE 276.

FIG. 3. HAPLOPHRAGRMIUM CANARIENSE D'ORBIGNY. SEE PAGE 277.

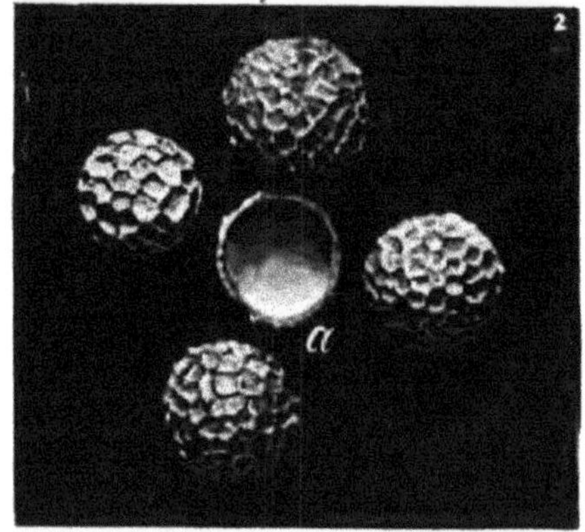

FIG. 1. HAPLOPHRAGMIUM GLOBIGERINIFORME PARKER & JONES. SEE PAGE 277.

FIG. 2. THURAMMINA FAVOSA NEW SPECIES. SEE PAGE 278.

a. Section.

FIG. 3. HAPLOSTICHE SOLDANII JONES & PARKER. SEE PAGE 277.

a. Longitudinal Section.

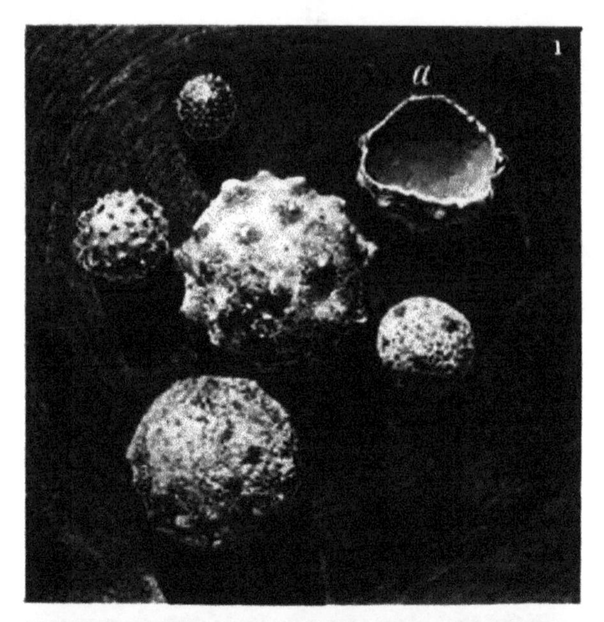

FIG. 1. THURAMMINA PAPILLATA BRADY. SEE PAGE 278.

a. Accidental Section

FIG. 2. THURAMMINA CARIOSA NEW SPECIES. SEE PAGE 278.

a. Section

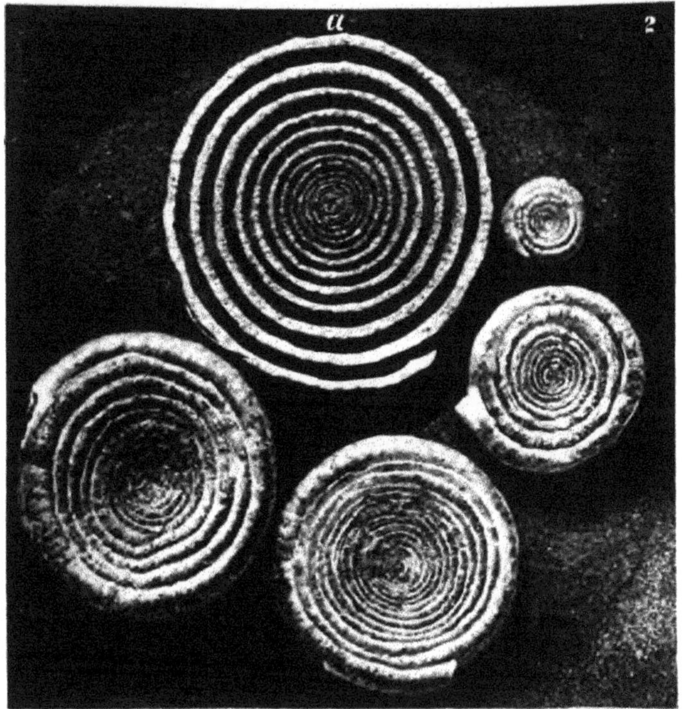

FIG. 1. AMMODISCUS TENUIS BRADY. SEE PAGE 279.

a. Section

FIG. 2. AMMODISCUS INCERTUS D'ORBIGNY. SEE PAGE 278.

a. Section

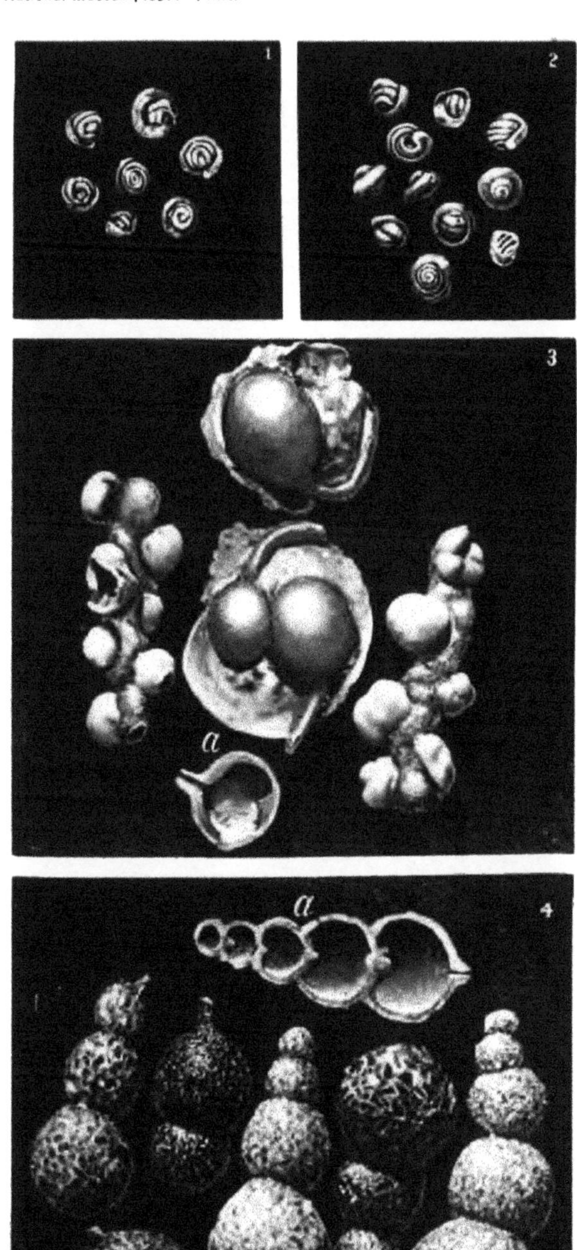

FIG. 1. AMMODISCUS GORDIALIS JONES & PARKER. SEE PAGE 279.
FIG. 2. AMMODISCUS CHAROIDES JONES & PARKER. SEE PAGE 279.
FIG. 3. WEBBINA CLAVATA JONES & PARKER. SEE PAGE 279.
a. Detached Specimen showing adherent face.
FIG. 4. HORMOSINA GLOBULIFERA BRADY. SEE PAGE 280.
a. Longitudinal Section.

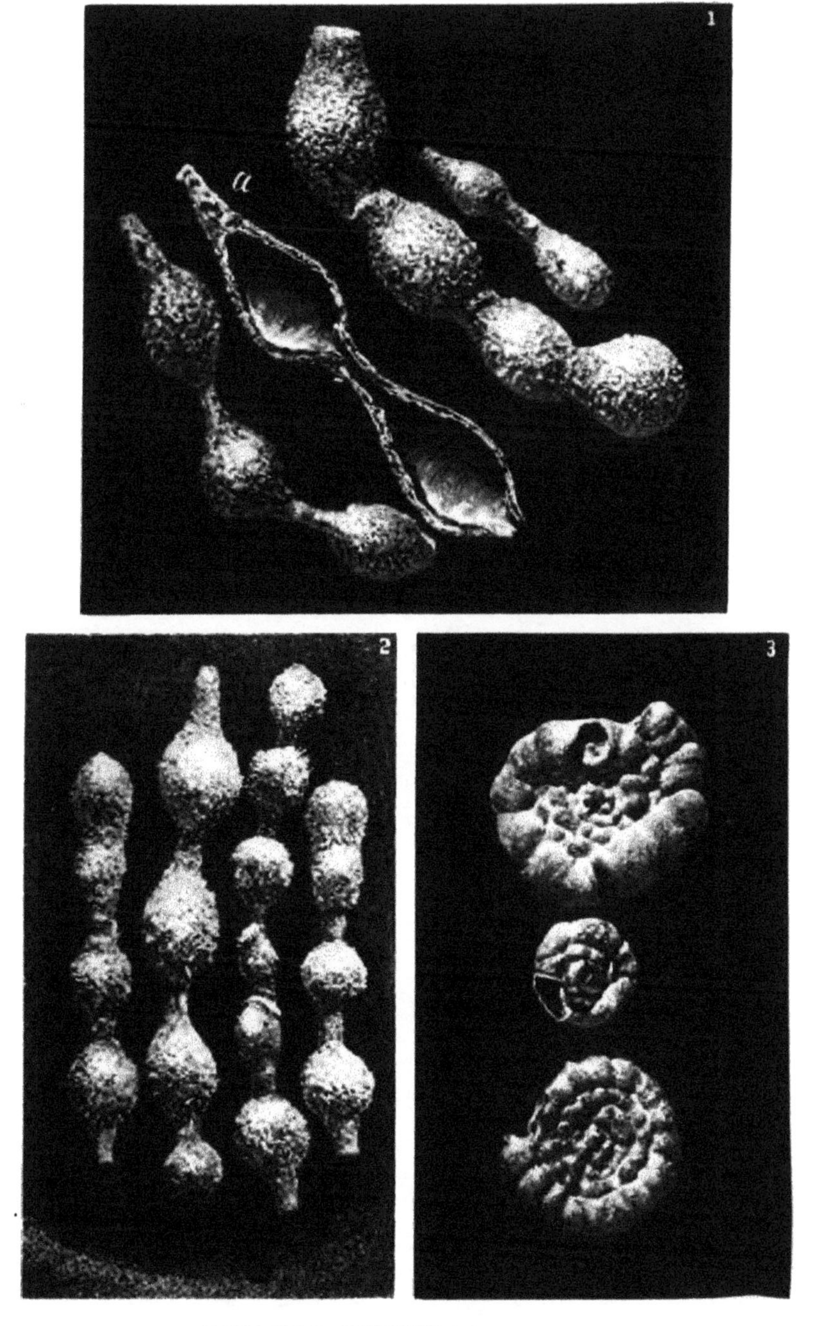

FIG. 1. HORMOSINA CARPENTERI BRADY. SEE PAGE 280.
a. Longitudinal Section.
FIG. 2. HORMOSINA OVICULA BRADY. SEE PAGE 280.
FIG. 3. TROCHAMMINA PROTEUS KARRER. SEE PAGE 281.

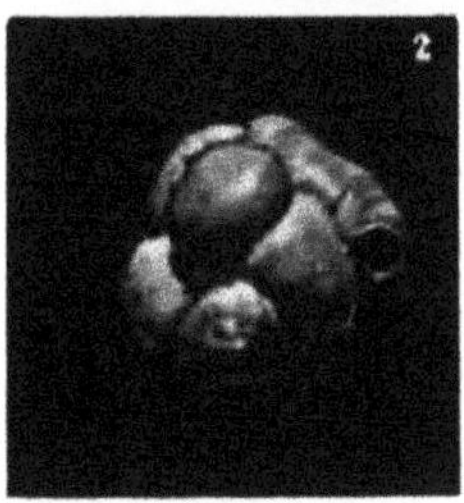

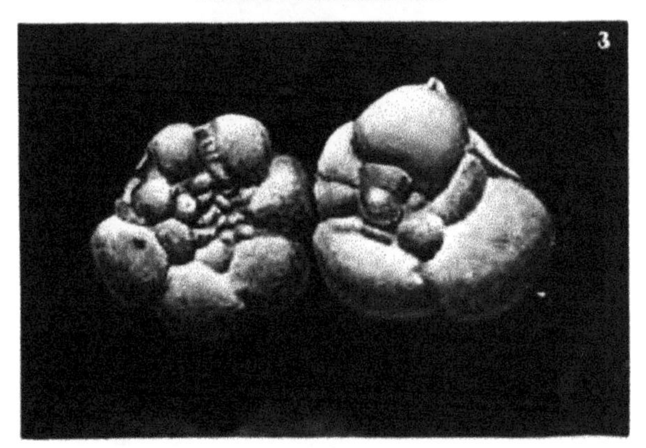

FIG. 1. TROCHAMMINA LITUIFORMIS BRADY. SEE PAGE 281.
FIG. 2. TROCHAMMINA CONGLOBATA BRADY. SEE PAGE 281.
FIG. 3. TROCHAMMINA CORONATA BRADY. SEE PAGE 281.

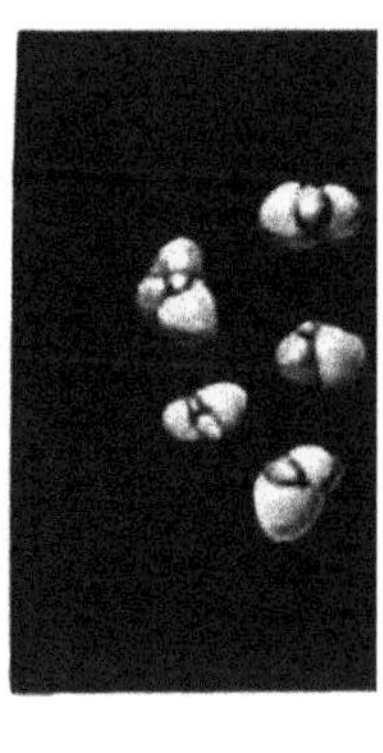

FIG. 1. TROCHAMMINA RINGENS BRADY. SEE PAGE 281.
FIG. 2. TROCHAMMINA PAUCILOCULATA BRADY. SEE PAGE 282.
FIG. 3. CYCLAMMINA CANCELLATA BRADY. SEE PAGE 282.
a. Section

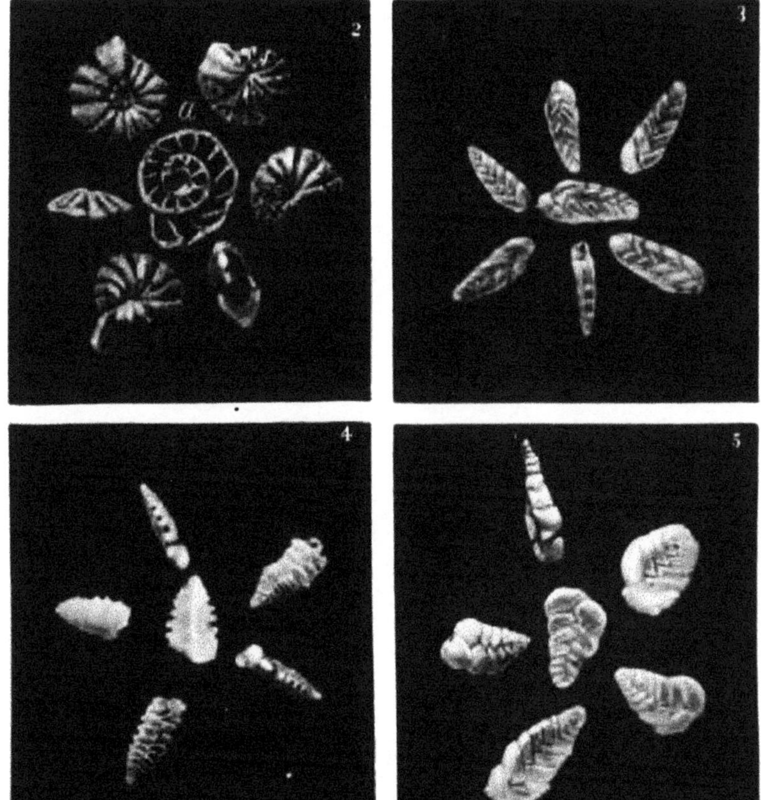

FIG. 1. CYCLAMMINA CANCELLATA BRADY. SMALL AND SMOOTH VARIETY. SEE PAGE 282.
a. Section.
FIG. 2. CYCLAMMINA PUSILLA BRADY. SEE PAGE 282.
a. Section.
FIG. 3. TEXTULARIA QUADRILATERA SCHWAGER. SEE PAGE 283.
FIG. 4. TEXTULARIA TRANSVERSARIA BRADY. SEE PAGE 283.
FIG. 5. TEXTULARIA CONCAVA KARRER. SEE PAGE 283.

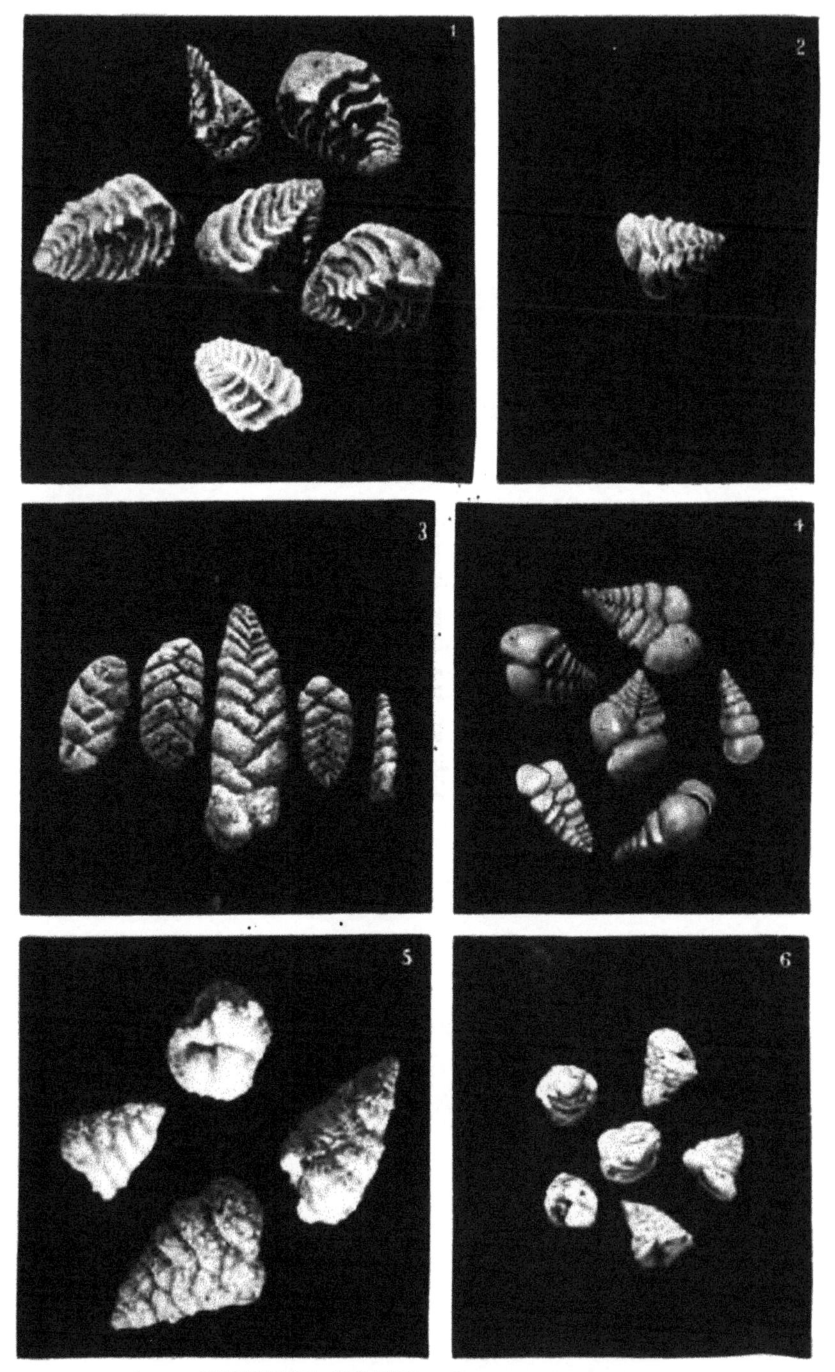

FIG. 1. TEXTULARIA CARINATA D'ORBIGNY. SEE PAGE 284.
FIG. 2. TEXTULARIA RUGOSA REUSS. SEE PAGE 284.
FIG. 3. TEXTULARIA LUCULENTA BRADY. SEE PAGE 284.
FIG. 4. TEXTULARIA AGGLUTINANS D'ORBIGNY. SEE PAGE 284.
FIG. 5. TEXTULARIA GRAMEN D'ORBIGNY. SEE PAGE 284.
FIG. 6. TEXTULARIA CONICA D'ORBIGNY. SEE PAGE 285.

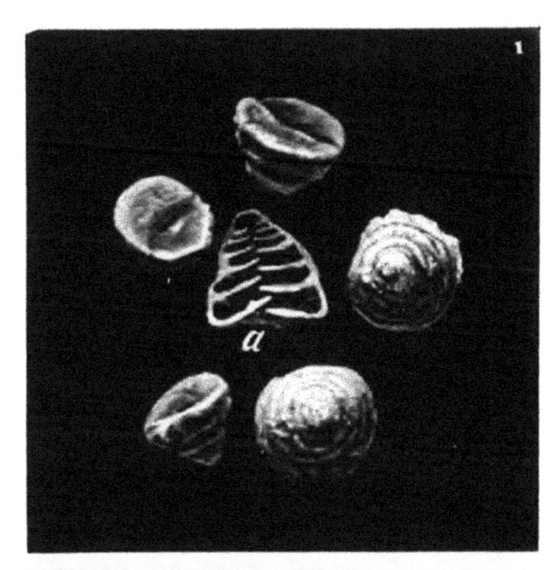

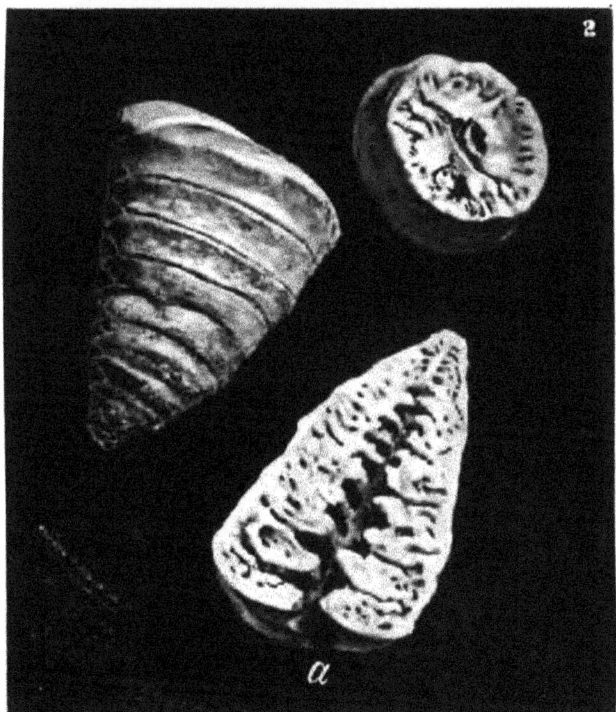

FIG. 1. TEXTULARIA TROCHUS D'ORBIGNY. SEE PAGE 285.
a. Longitudinal Section

FIG. 2. TEXTULARIA BARRETTII JONES & PARKER. SEE PAGE 285.
a. Longitudinal Section

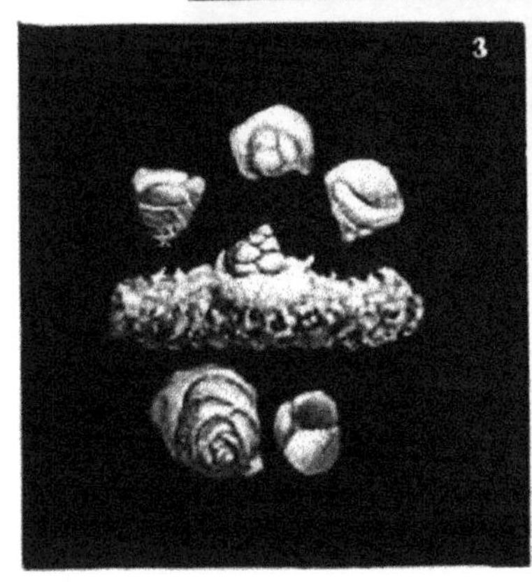

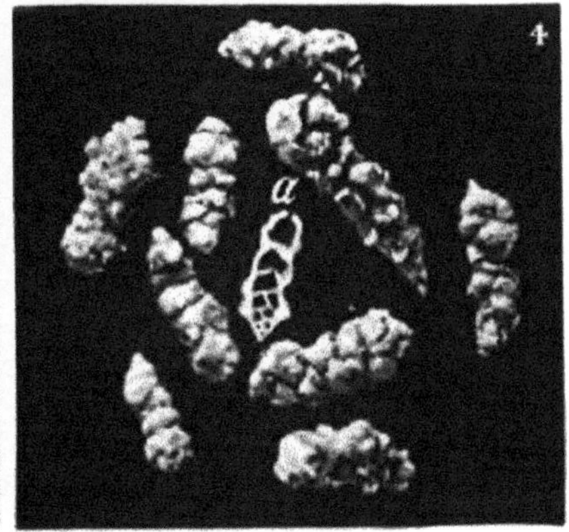

FIG. 1. VERNEUILINA PYGMÆA EGGER. SEE PAGE 285.
FIG. 2. VERNEUILINA PROPINQUA BRADY. SEE PAGE 285.
FIG. 3. VALVULINA CONICA PARKER & JONES. SEE PAGE 286.
FIG. 4. BIGENERINA NODOSARIA D'ORBIGNY. SEE PAGE 286.
a. Longitudinal Section.

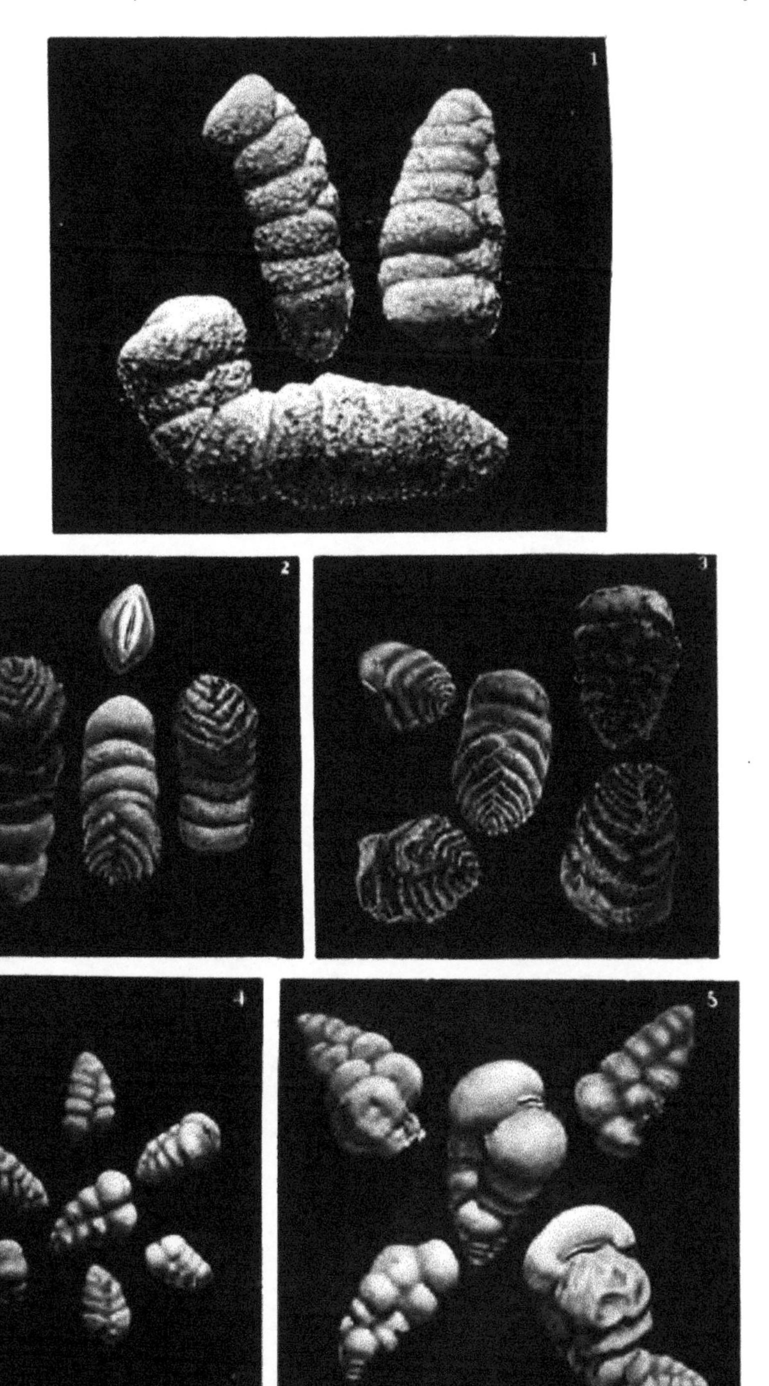

FIG. 1. BIGENERINA ROBUSTA BRADY. SEE PAGE 286.
FIG. 2. BIGENERINA PENNATULA BATSCH. SEE PAGE 287.
FIG. 3. BIGENERINA CAPREOLUS D'ORBIGNY. SEE PAGE 286.
FIG. 4. GAUDRYINA PUPOIDES D'ORBIGNY. SEE PAGE 287.
FIG. 5. GAUDRYINA BACCATA SCHWAGER. SEE PAGE 287.

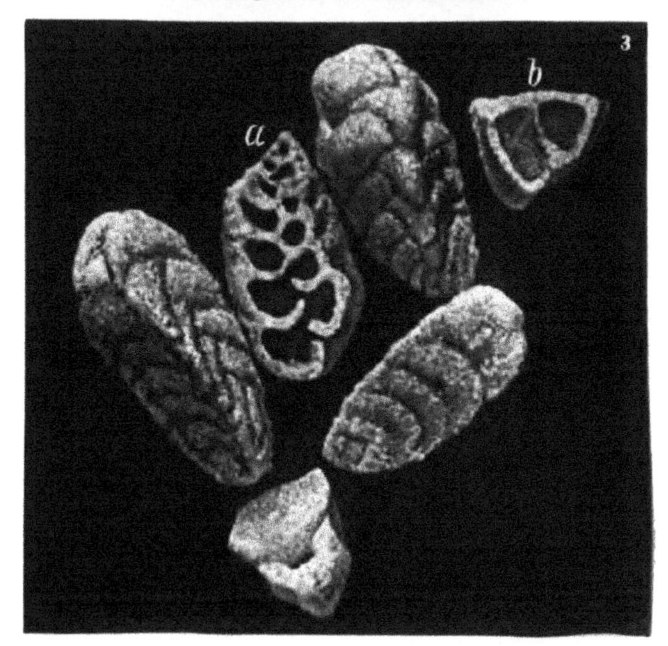

FIG. 1. GAUDRYINA SUBROTUNDATA SCHWAGER. SEE PAGE 287.
FIG. 2. GAUDRYINA FILIFORMIS BERTHELIN. SEE PAGE 287.
FIG. 3. GAUDRYINA RUGOSA D'ORBIGNY. SEE PAGE 288.
a. Longitudinal Section.

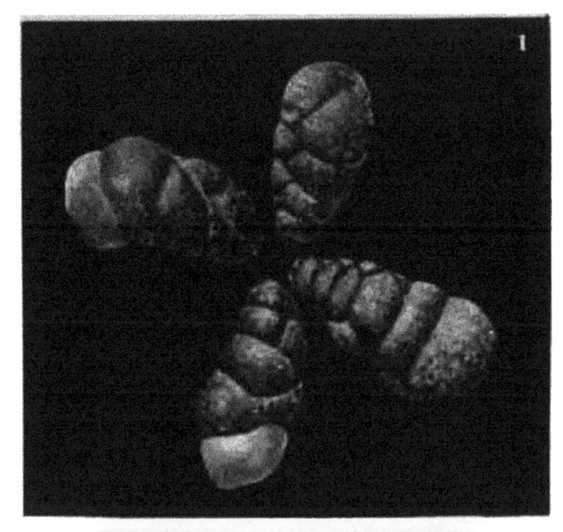

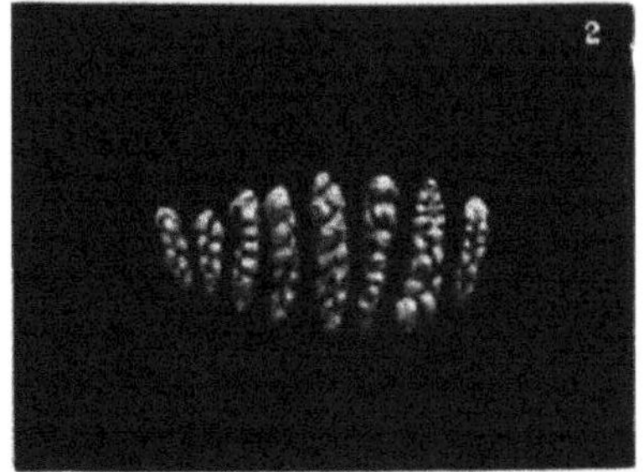

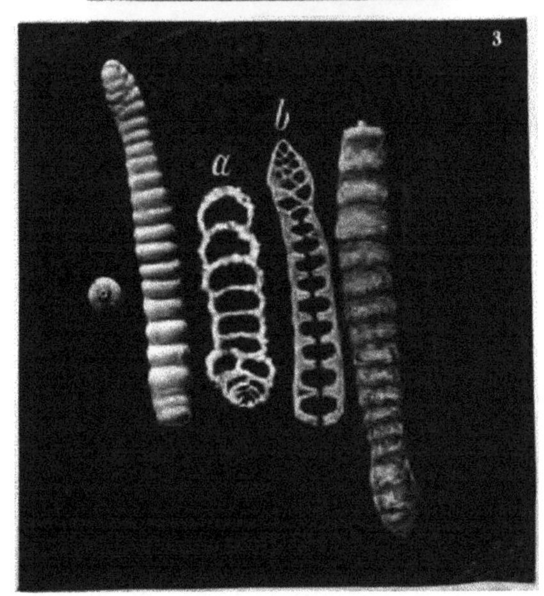

FIG. 1. GAUDRYINA SCABRA BRADY. SEE PAGE 288.
FIG. 2. GAUDRYINA SIPHONELLA REUSS. SEE PAGE 288.
FIG. 3. CLAVULINA COMMUNIS D'ORBIGNY. SEE PAGE 288.
a, b. Longitudinal Sections.

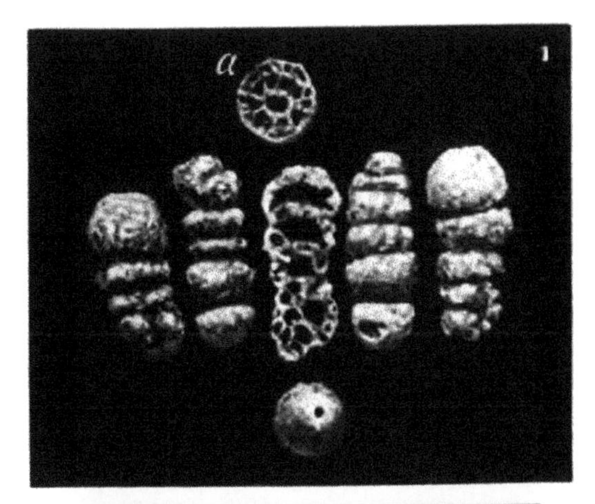

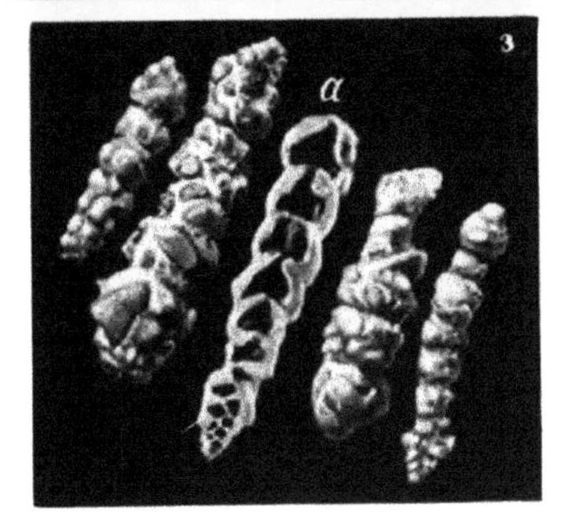

FIG. 1. CLAVULINA EOCÆNA GÜMBEL. SEE PAGE 289.

a. Transverse Section.

FIG. 2. CLAVULINA PARISIENSIS D'ORBIGNY. SEE PAGE 289.

FIG. 3. CLAVULINA PARISIENSIS D'ORBIGNY. (VAR COARSE CORAL SAND.) SEE PAGE 289

a. Longitudinal Section

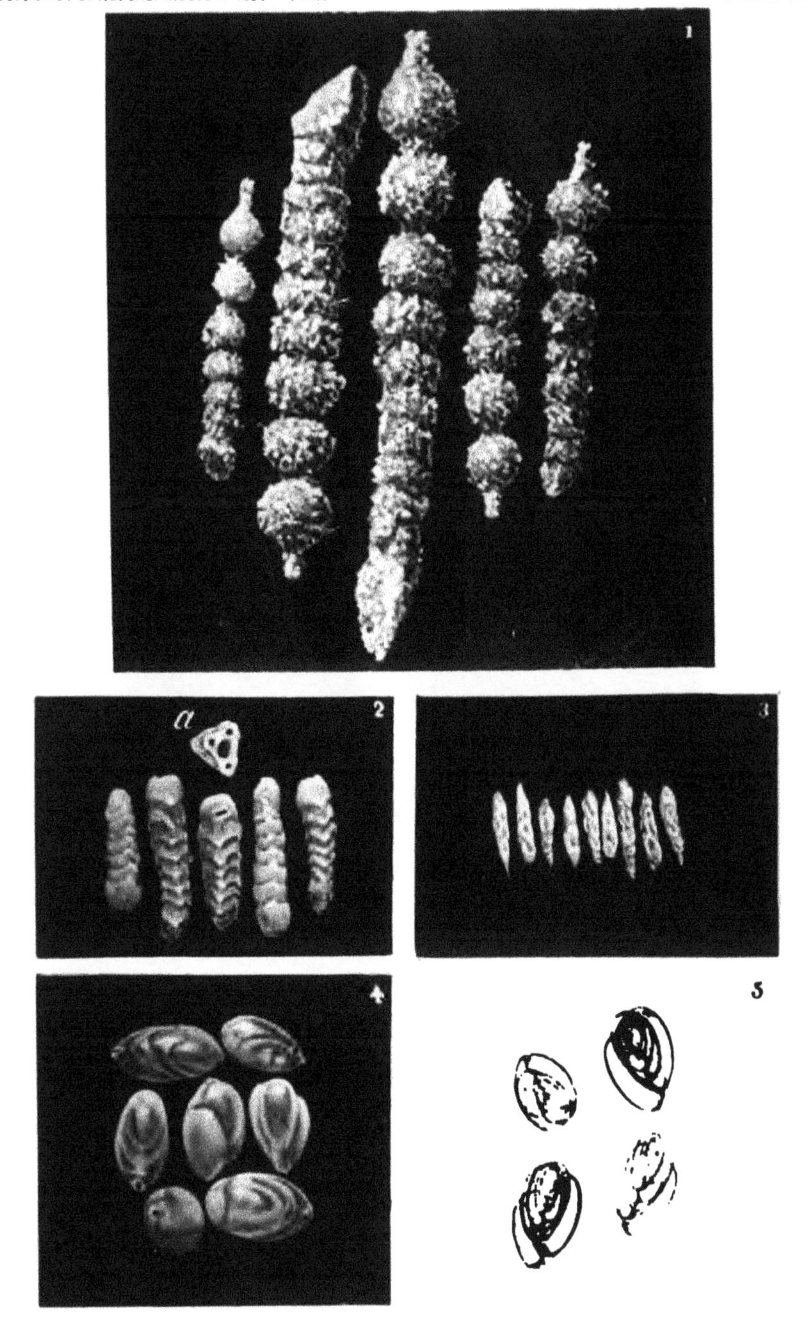

FIG. 1. CLAVULINA PARISIENSIS D'ORBIGNY, VAR. HUMILIS BRADY. SEE PAGE 289.
FIG. 2. CLAVULINA ANGULARIS D'ORBIGNY. SEE PAGE 289.
FIG. 3. BULIMINA ELEGANS D'ORBIGNY. SEE PAGE 290.
FIG. 4. BULIMINA PYRULA D'ORBIGNY. SEE PAGE 290.
FIG. 5. BULIMINA PYRULA D'ORBIGNY. (BY TRANSMITTED LIGHT.)

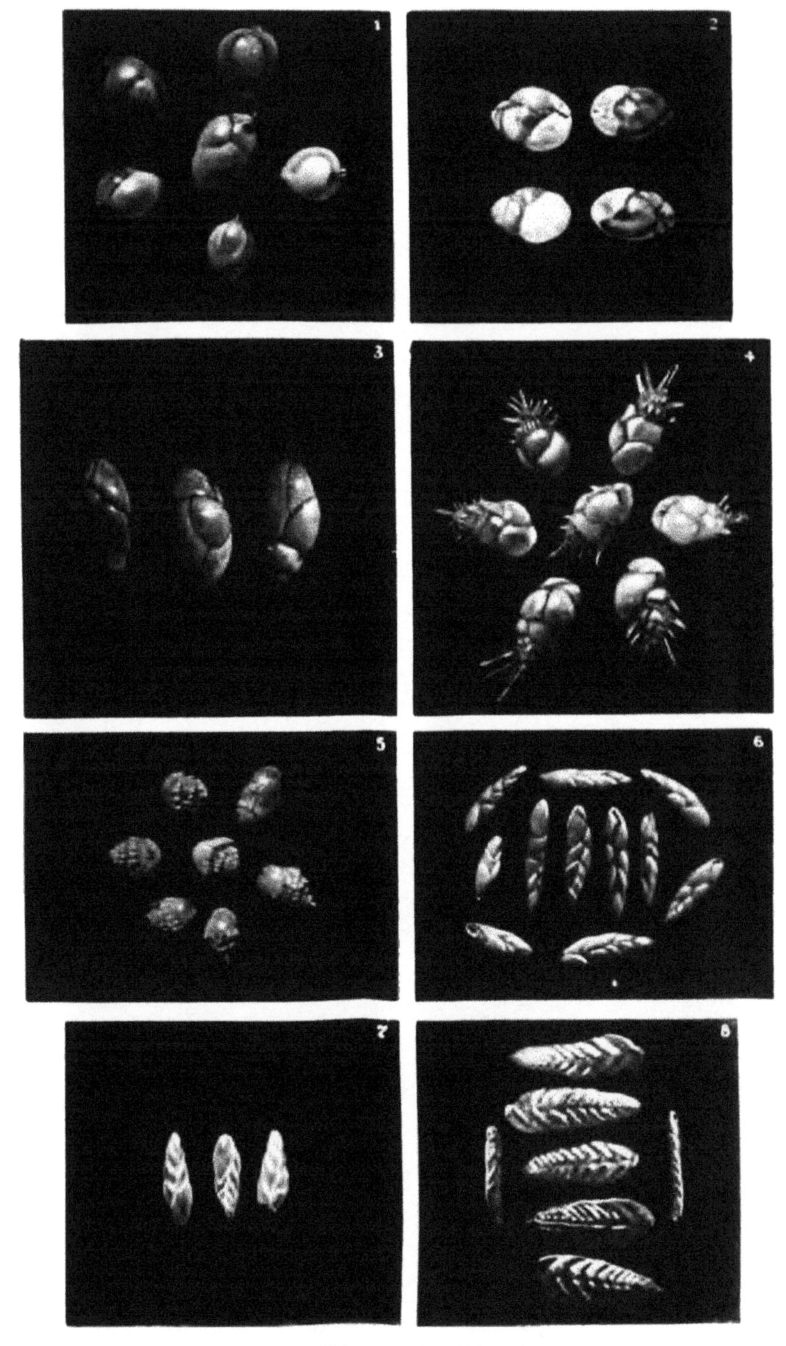

FIG. 1. BULIMINA PYRULA, VAR. SPINESCENS BRADY. SEE PAGE 290.
FIG. 2. BULIMINA AFFINIS D'ORBIGNY. SEE PAGE 290.
FIG. 3. BULIMINA PUPOIDES D'ORBIGNY. SEE PAGE 290.
FIG. 4. BULIMINA ACULEATA D'ORBIGNY. SEE PAGE 291.
FIG. 5. BULIMINA INFLATA SEGUENZA. SEE PAGE 291.
FIG. 6. VIRGULINA SCHREIBERSIANA CZJZEK. SEE PAGE 291.
FIG. 7. VIRGULINA SUBSQUAMOSA EGGER. SEE PAGE 291.
FIG. 8. *BOLIVINA* ÆNARIENSIS COSTA. SEE PAGE 292.

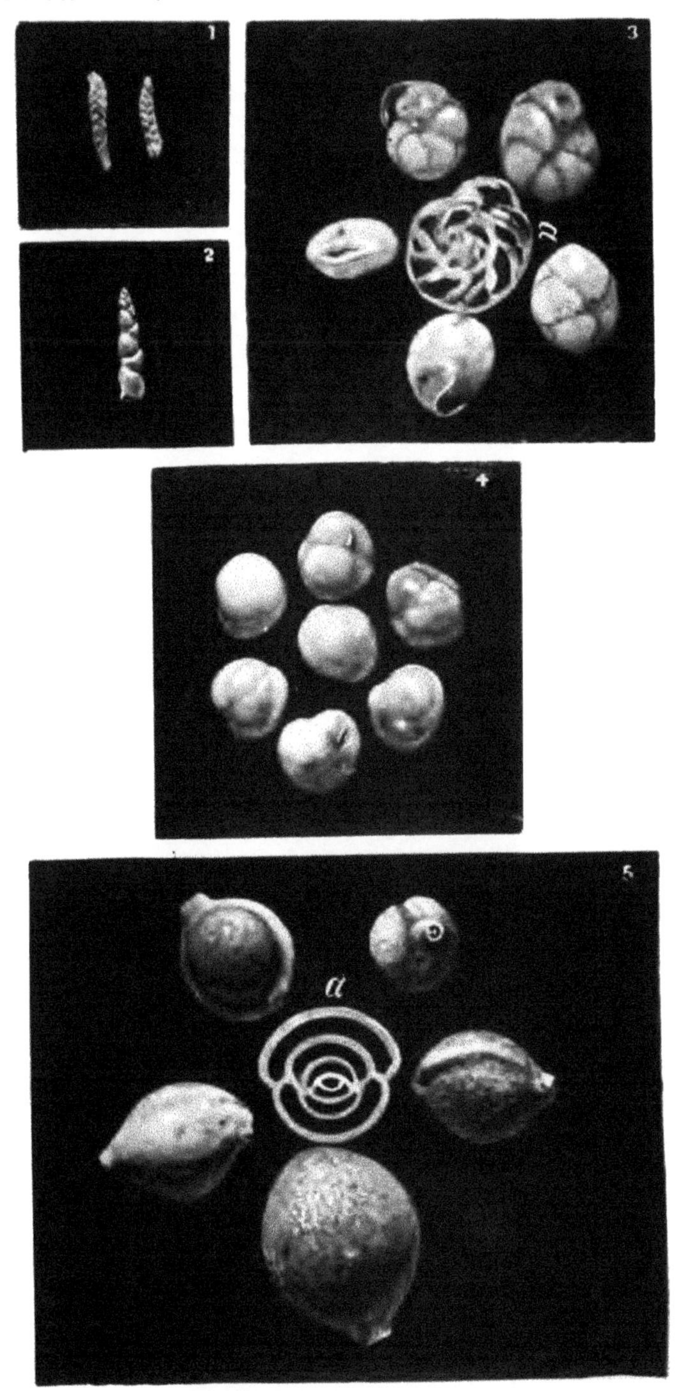

FIG. 1. BOLIVINA PUNCTATA D'ORBIGNY. SEE PAGE 292.
FIG. 2. BOLIVINA PORRECTA BRADY. SEE PAGE 292.
FIG. 3. CASSIDULINA CRASSA D'ORBIGNY. SEE PAGE 292.
FIG. 4. CASSIDULINA SUBGLOBOSA BRADY. SEE PAGE 293.
FIG. 5. BILOCULINA BULLOIDES D'ORBIGNY. SEE PAGE 293.
a. Transverse Section.

FIG. 1. BILOCULINA TUBULOSA COSTA. SEE PAGE 293.
FIG. 2. BILOCULINA RINGENS LAMARCK. SEE PAGE 294.
a. Transverse Section.
FIG. 3. BILOCULINA COMATA BRADY. SEE PAGE 294.
FIG. 4. BILOCULINA ELONGATA EHRENBERG. SEE PAGE 294.

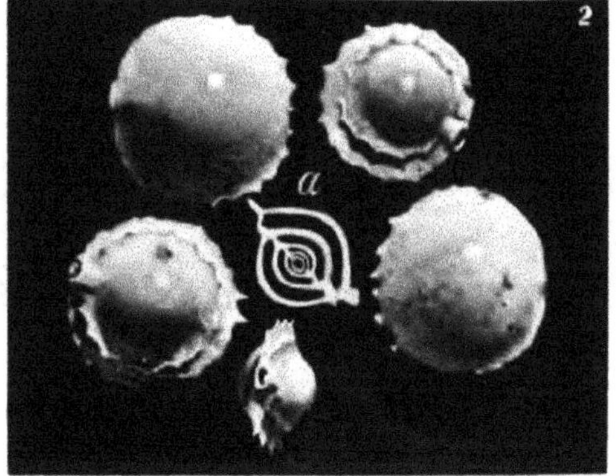

FIG. 1. BILOCULINA DEPRESSA D'ORBIGNY. SEE PAGE 294.

a. Transverse Section.

FIG. 2. BILOCULINA DEPRESSA, VAR. SERRATA BRADY. SEE PAGE 294.

a. Transverse Section

FIG. 3. BILOCULINA DEHISCENS NEW SPECIES. SEE PAGE 295.

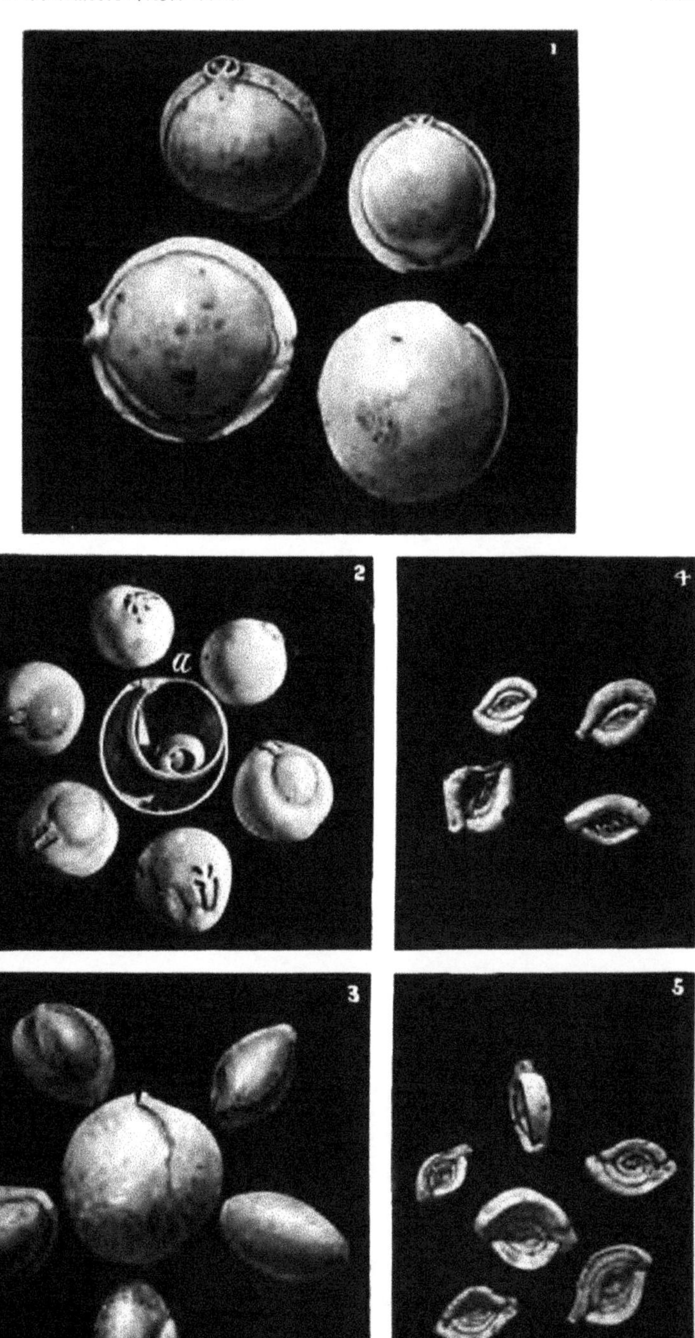

FIG. 1. BILOCULINA LÆVIS DEFRANCE. SEE PAGE 295.
FIG. 2. BILOCULINA SPHÆRA D'OBRIGNY. SEE PAGE 295.
a. Section.
FIG. 3. BILOCULINA IRREGULARIS D'ORBIGNY. SEE PAGE 295.
FIG. 4. SPIROLOCULINA NITIDA D'ORBIGNY. SEE PAGE 296.
FIG. 5. *SPIROLOCULINA* EXCAVATA D'ORBIGNY. SEE PAGE 296.

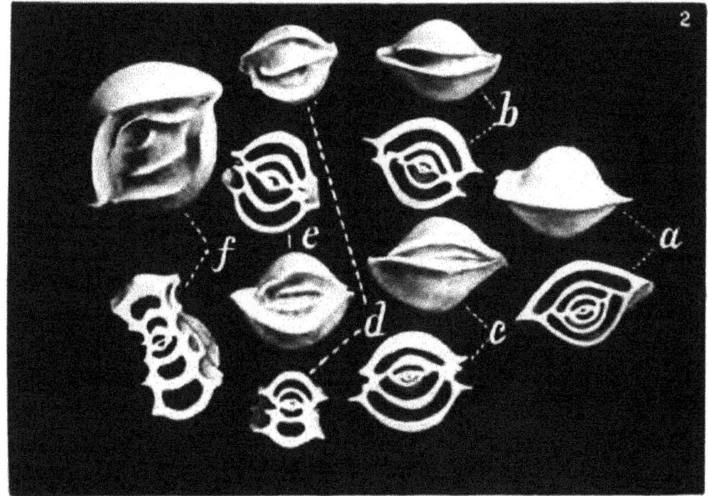

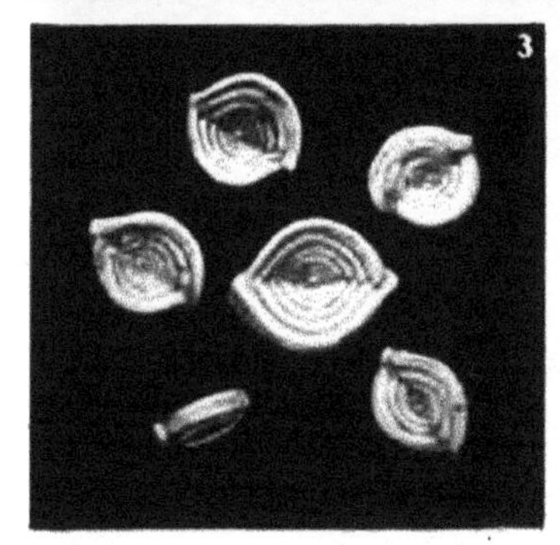

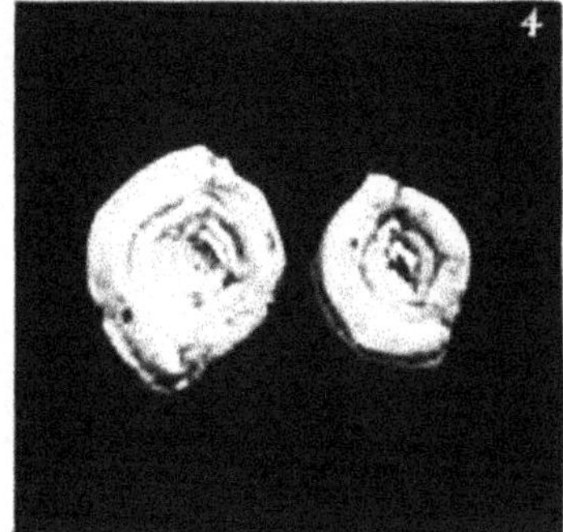

FIG. 1. SPIROLOCULINA ROBUSTA BRADY. SEE PAGE 296.

a. Horizontal Section. *b.* Transverse Section.

FIG. 2. SPIROLOCULINA ROBUSTA BRADY. (TRANSITION STAGES FROM BILOCULINA COMPRESSA.) SEE PAGE 296.

FIG. 3. SPIROLOCULINA LIMBATA D'ORBIGNY. SEE PAGE 296.

FIG. 4. SPIROLOCULINA PLANULATA LAMARCK. SEE PAGE 297.

FIG. 1. SPIROLOCULINA ARENARIA BRADY. SEE PAGE 297.
FIG. 2. MILIOLINA SEMINULUM LINNÆUS. SEE PAGE 297.
FIG. 3. MILIOLINA OBLONGA MONTAGU. SEE PAGE 297.
FIG. 4. MILIOLINA CUVIERANA D'ORBIGNY. SEE PAGE 298.
FIG. 5. MILIOLINA GRACILIS D'ORBIGNY. SEE PAGE 297.
FIG. 6. MILIOLINA AUBERIANA D'ORBIGNY. SEE PAGE 298.
a. Transverse Section

FIG. 1. MILIOLINA BUCCULENTA BRADY. SEE PAGE 299.
FIG. 2. MILIOLINA INSIGNIS BRADY. SEE PAGE 299.
a. Transverse Section.
FIG. 3. MILIOLINA LABIOSA D'ORBIGNY. SEE PAGE 299.
FIG. 4. MILIOLINA UNDOSA KARRER. SEE PAGE 300.

FIG. 1. MILIOLINA ANGULARIS NEW SPECIES. SEE PAGE 300.
FIG. 2. MILIOLINA BICORNIS WALKER & JACOB. SEE PAGE 300.
FIG. 3. MILIOLINA LINNAEANA D'ORBIGNY. SEE PAGE 300.
FIG. 4. MILIOLINA PULCHELLA D'ORBIGNY. SEE PAGE 301.
FIG. 5. MILIOLINA RETICULATA D'ORBIGNY. SEE PAGE 301.
FIG. 6. MILIOLINA SEPARANS BRADY. SEE PAGE 300.

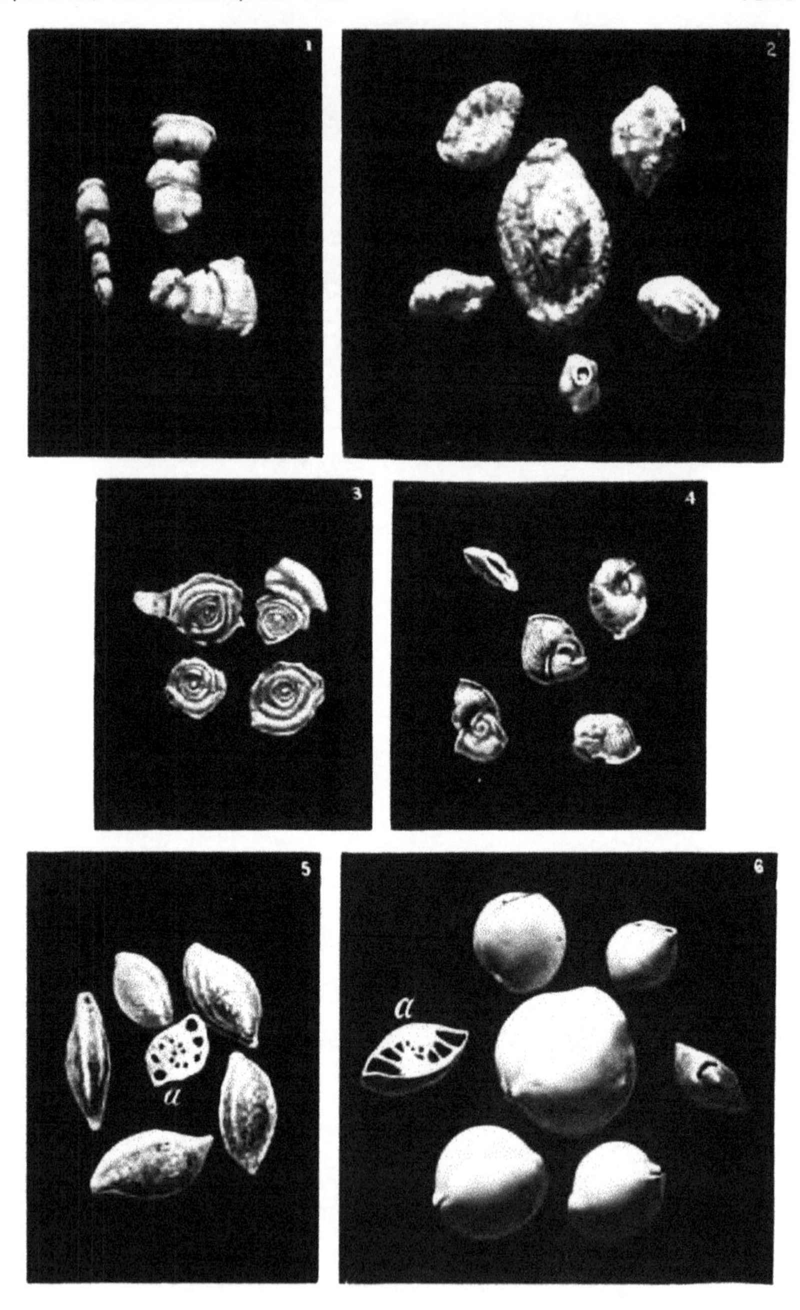

FIG. 1. ARTICULINA SAGRA D'ORBIGNY. SEE PAGE 301.
FIG. 2. MILIOLINA AGGLUTINANS D'ORBIGNY. SEE PAGE 301.
FIG. 3. OPHTHALMIDIUM INCONSTANS BRADY. SEE PAGE 302.
FIG. 4. VERTEBRALINA INSIGNIS BRADY. SEE PAGE 302.
FIG. 5. PLANISPIRINA CELATA COSTA. SEE PAGE 303.
a. Transverse Section.
FIG. 6. PLANISPIRINA SIGMOIDEA BRADY. SEE PAGE 302.
a. Transverse Section.

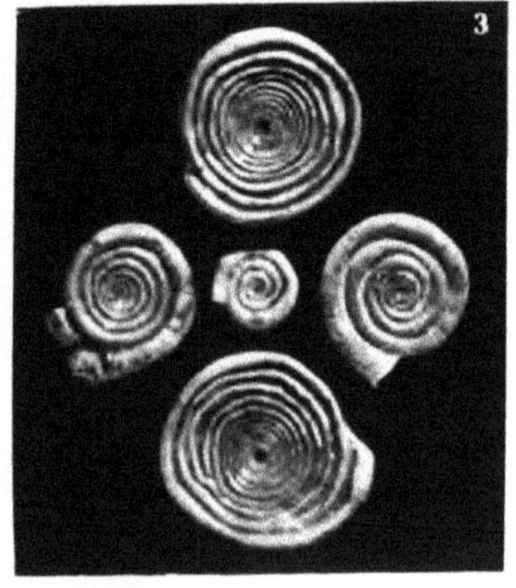

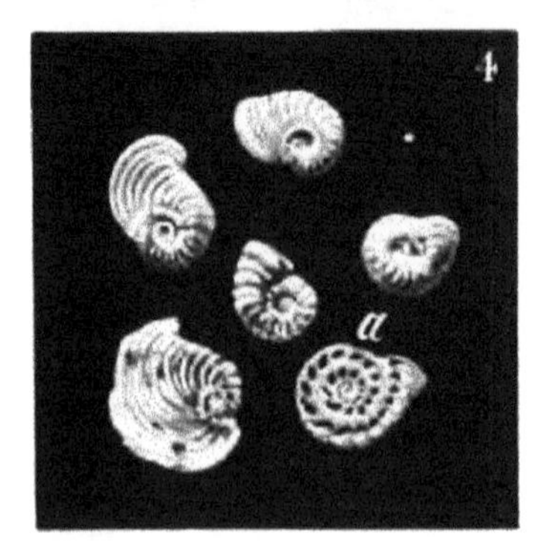

FIG. 1. CORNUSPIRA FOLIACEA PHILIPPI. SEE PAGE 303.
FIG. 2. CORNUSPIRA CARINATA COSTA. SEE PAGE 303.
FIG. 3. CORNUSPIRA INVOLVENS REUSS. SEE PAGE 303.
FIG. 4. PENEROPLIS PERTUSUS FÖRSKAL. SEE PAGE 304.
a Horizontal Section

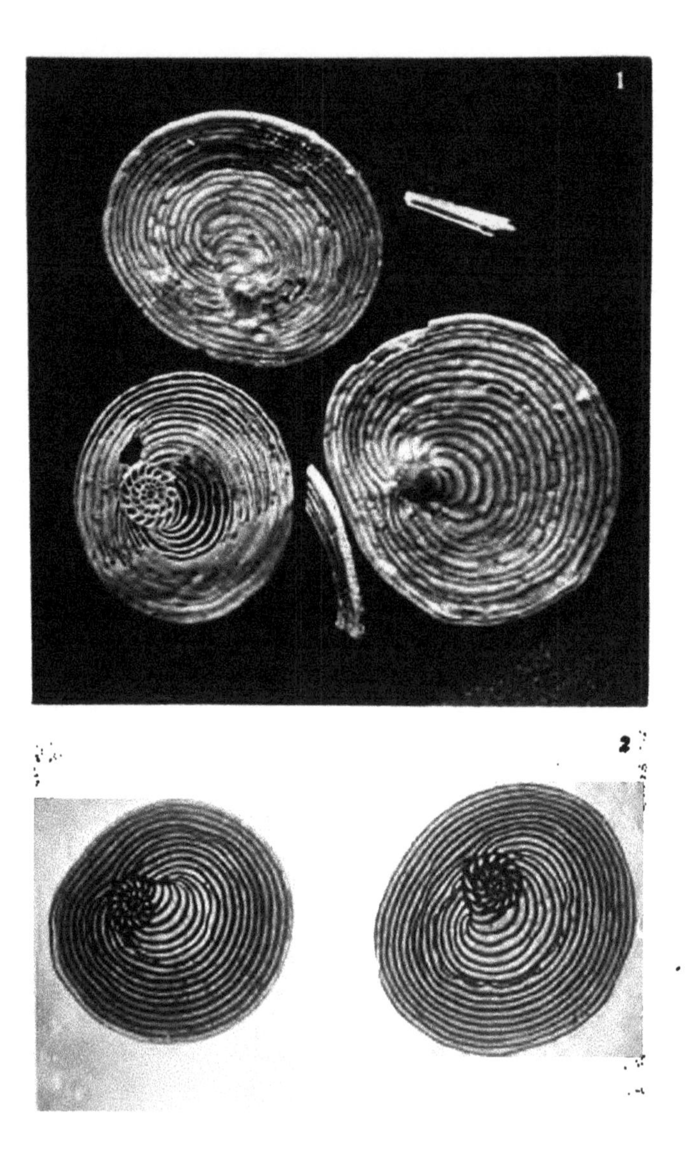

FIG. 1. PENEROPLIS PERTUSUS FÖRSKAL, VAR. DISCOIDEUS, NEW. SEE PAGE 804.
a. Incomplete Section

FIG. 2. PENEROPLIS PERTUSUS FÖRSKAL. (BY TRANSMITTED LIGHT.) SEE PAGE 804.

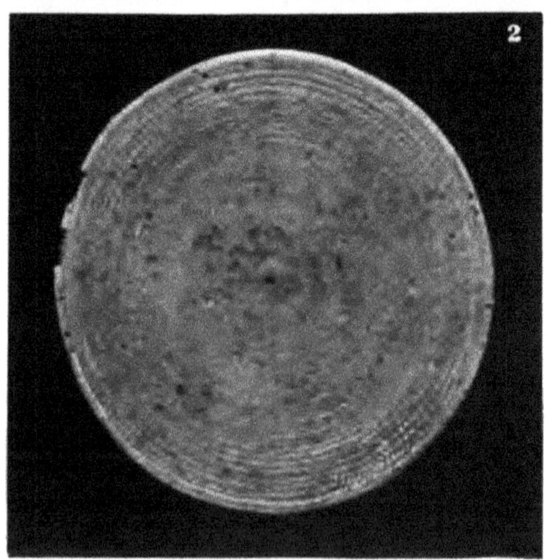

FIG. 1. ORBICULINA ADUNCA FICHTEL & MOLL. SEE PAGE 304.
FIG. 2. ORBITOLITES MARGINALIS LAMARCK. SEE PAGE 304.

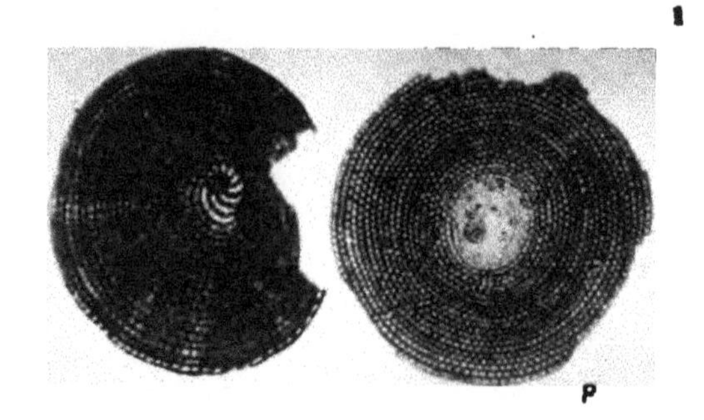

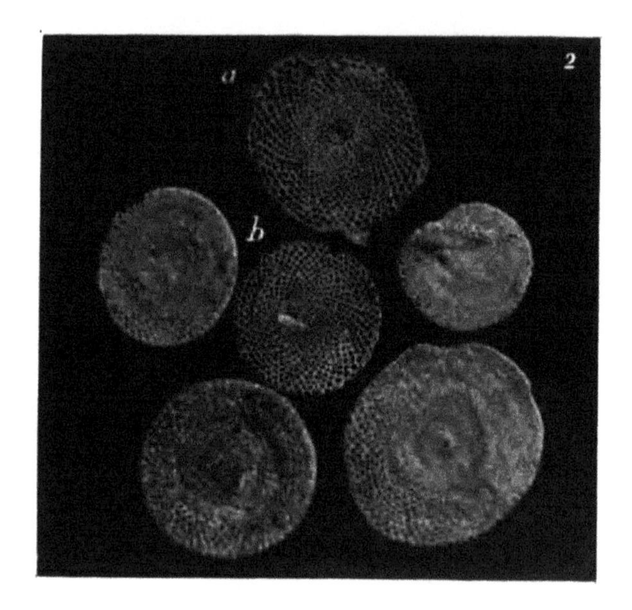

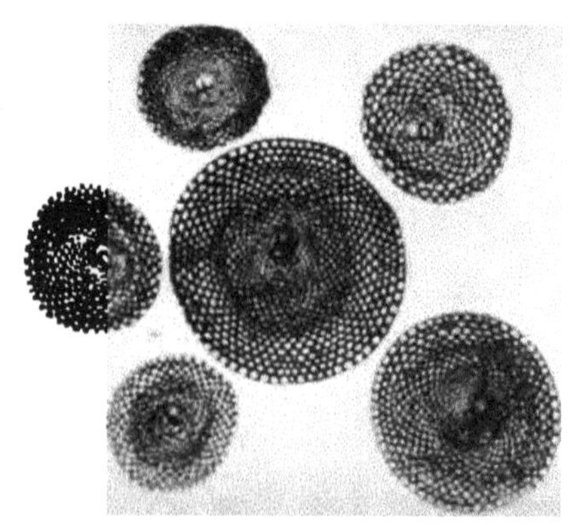

3. 1. ORBITOLITES MARGINALIS LAMARCK. BY TRANSMITTED LIGHT. SEE PAGE 304.

a. : or

G. 2. ORBITOLITES DUPLEX CARPENTER. SEE PAGE 305

a, b Sections.

3. 3. *ORBITOLITES DUPLEX* CARPENTER. (BY TRANSMITTED LIGHT.

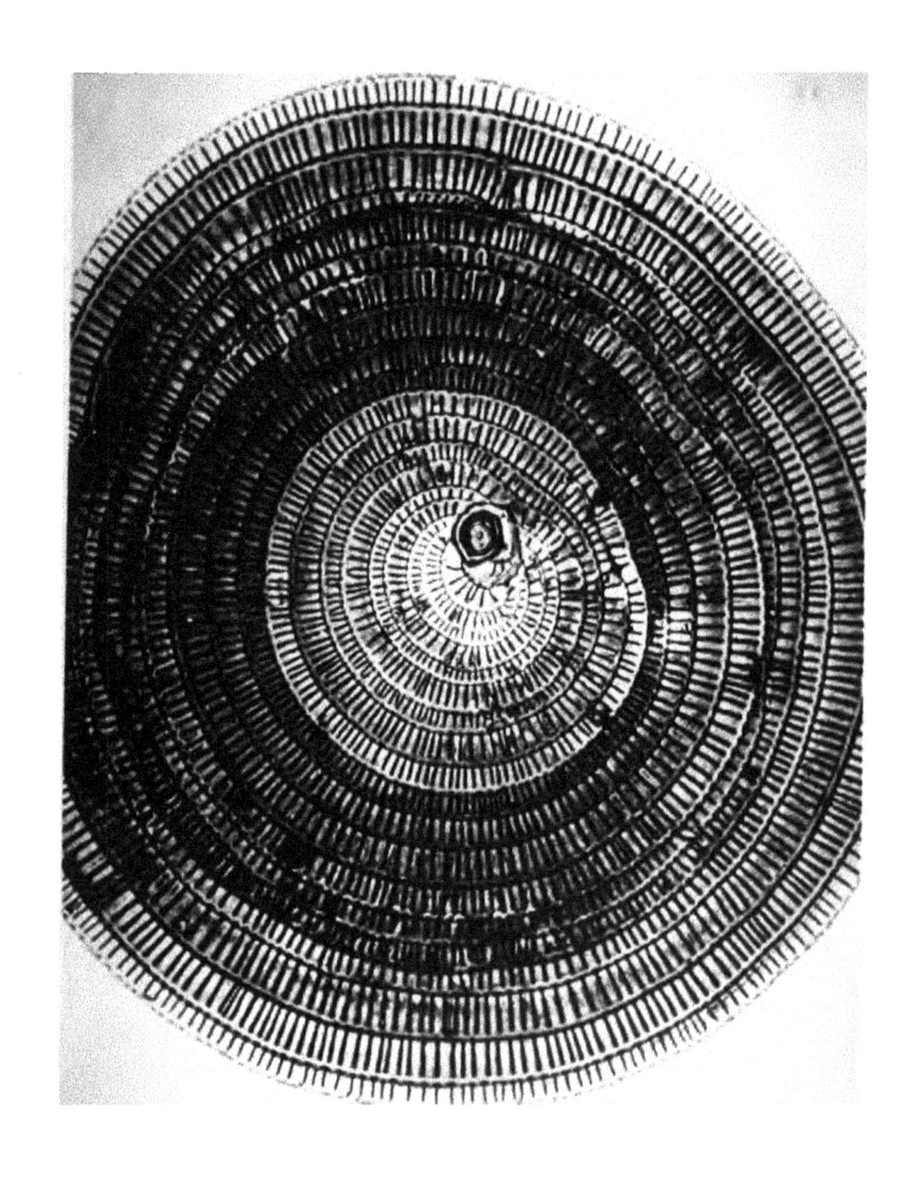

ORBITOLITES TENUISSIMA (CARPENTER), BY TRANSMITTED LIGHT.

[illegible]

The shaded portion of the [illegible] the anima

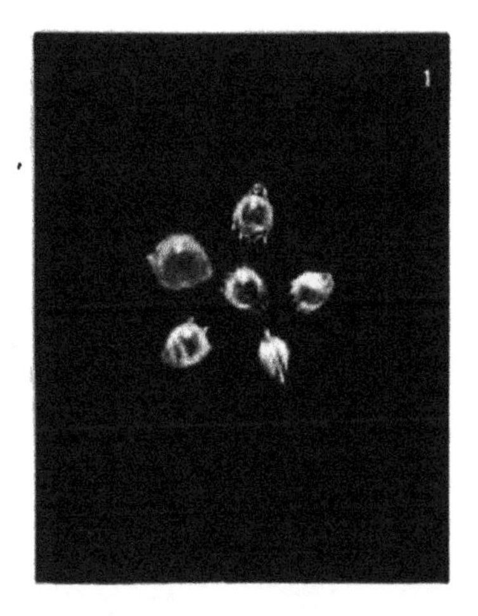

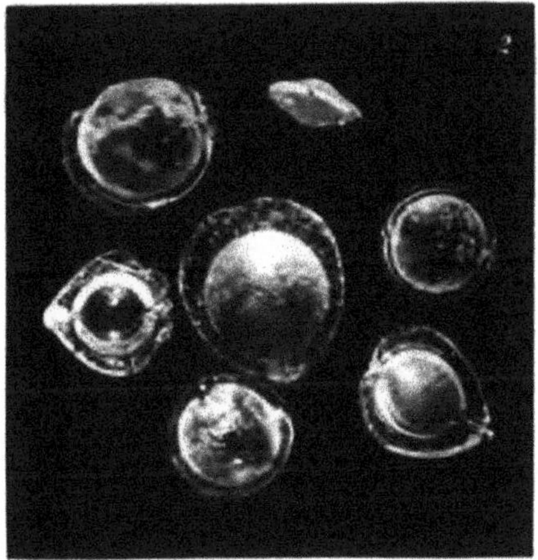

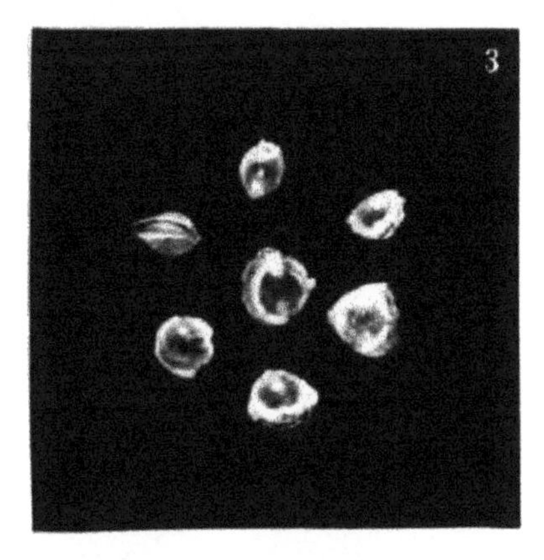

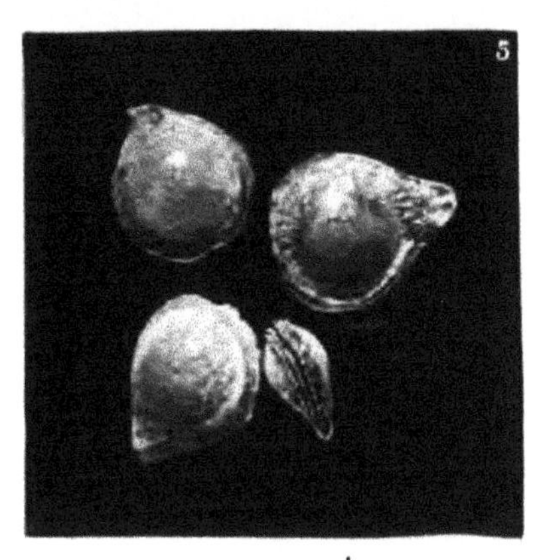

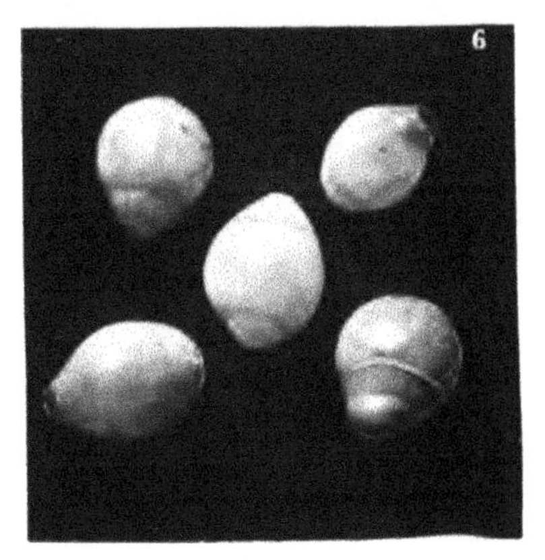

FIG. 1. LAGENA STAPHYLLEARIA SCHWAGER. SEE PAGE 307.

FIG. 2. LAGENA MARGINATA WALKER & BOYS. SEE PAGE 307.

FIG. 3. LAGENA CASTANEA NEW SPECIES. SEE PAGE 307.

FIG. 4. LAGENA ORBIGNYANA SEGUENZA. SEE PAGE 308.

FIG. 5. LAGENA CASTRENSIS SCHWAGER. SEE PAGE 308.

FIG. 6. NODOSARIA ROTUNDATA REUSS. SEE PAGE 308.

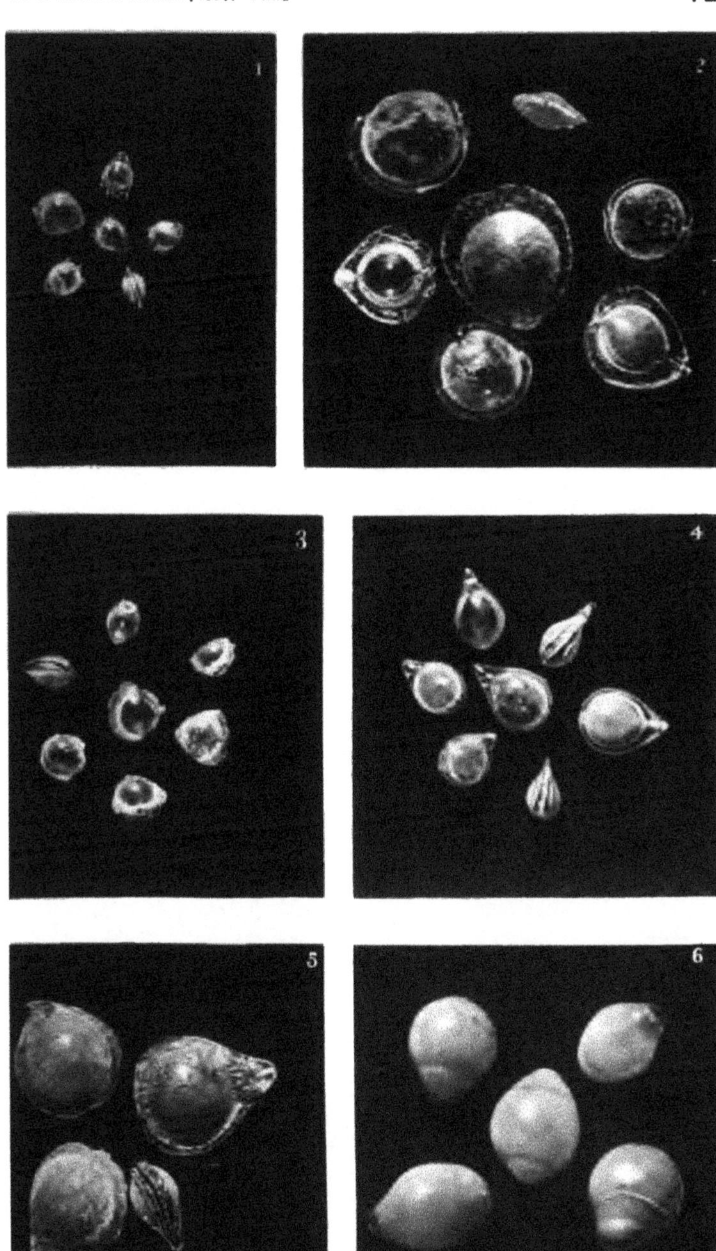

FIG. 1. LAGENA STAPHYLLEARIA SCHWAGER. SEE PAGE 307.
FIG. 2. LAGENA MARGINATA WALKER & BOYS. SEE PAGE 307.
FIG. 3. LAGENA CASTANEA NEW SPECIES. SEE PAGE 307.
FIG. 4. LAGENA ORBIGNYANA SEGUENZA. SEE PAGE 308.
FIG. 5. *LAGENA* CASTRENSIS SCHWAGER. SEE PAGE 308.
FIG. 6. *NODOSARIA* ROTUNDATA REUSS. SEE PAGE 308.

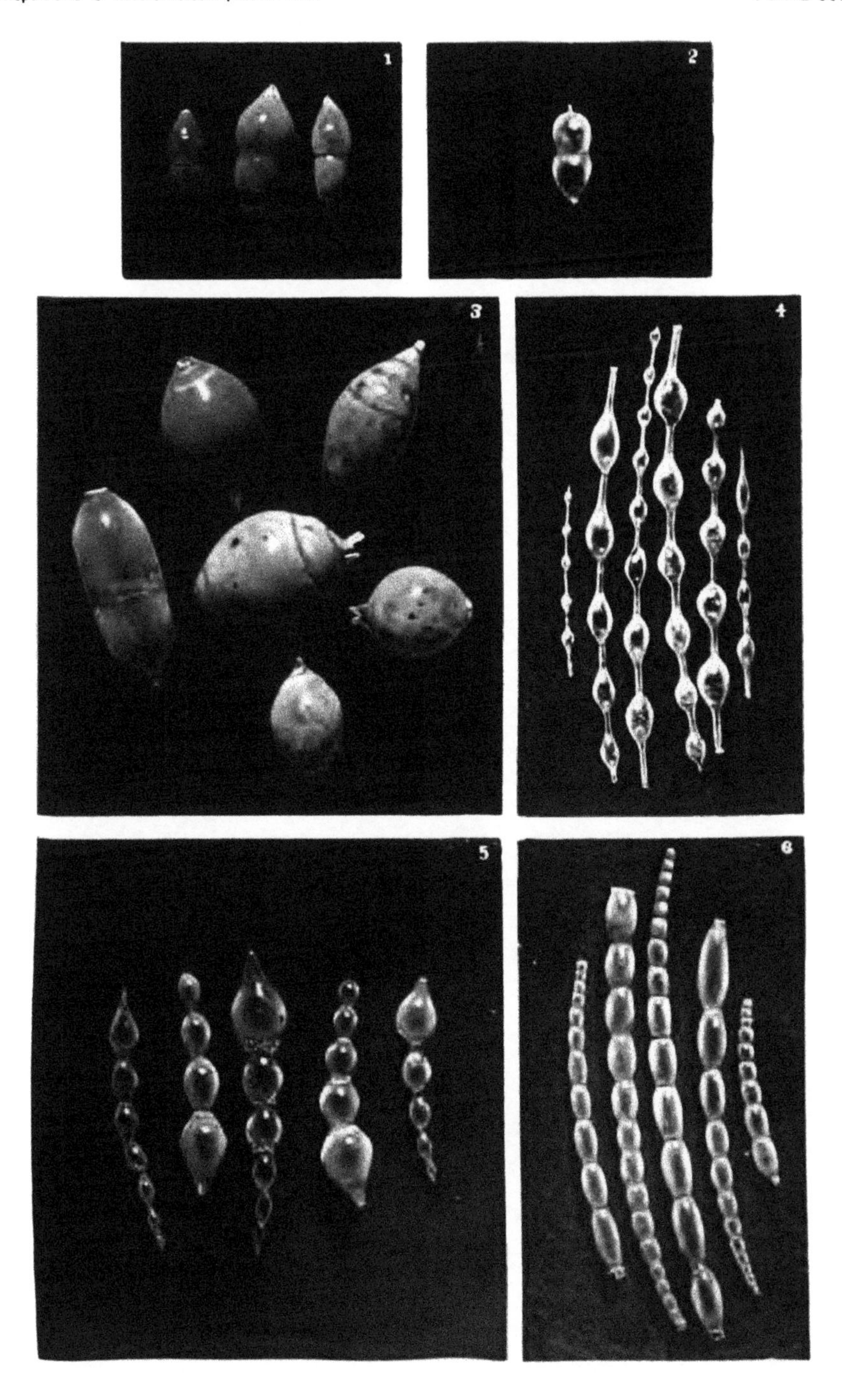

FIG. 1. NODOSARIA RADICULA LINNÆUS. SEE PAGE 309.
FIG. 2. NODOSARIA SIMPLEX SYLVESTRI. SEE PAGE 309.
FIG. 3. NODOSARIA LÆVIGATA NILSSON. SEE PAGE 308.
FIG. 4. NODOSARIA PYRULA D'ORBIGNY. SEE PAGE 309.
FIG. 5. NODOSARIA FARCIMEN SOLDANI. SEE PAGE 309.
FIG. 6. NODOSARIA FILIFORMIS D'ORBIGNY. SEE PAGE 310.

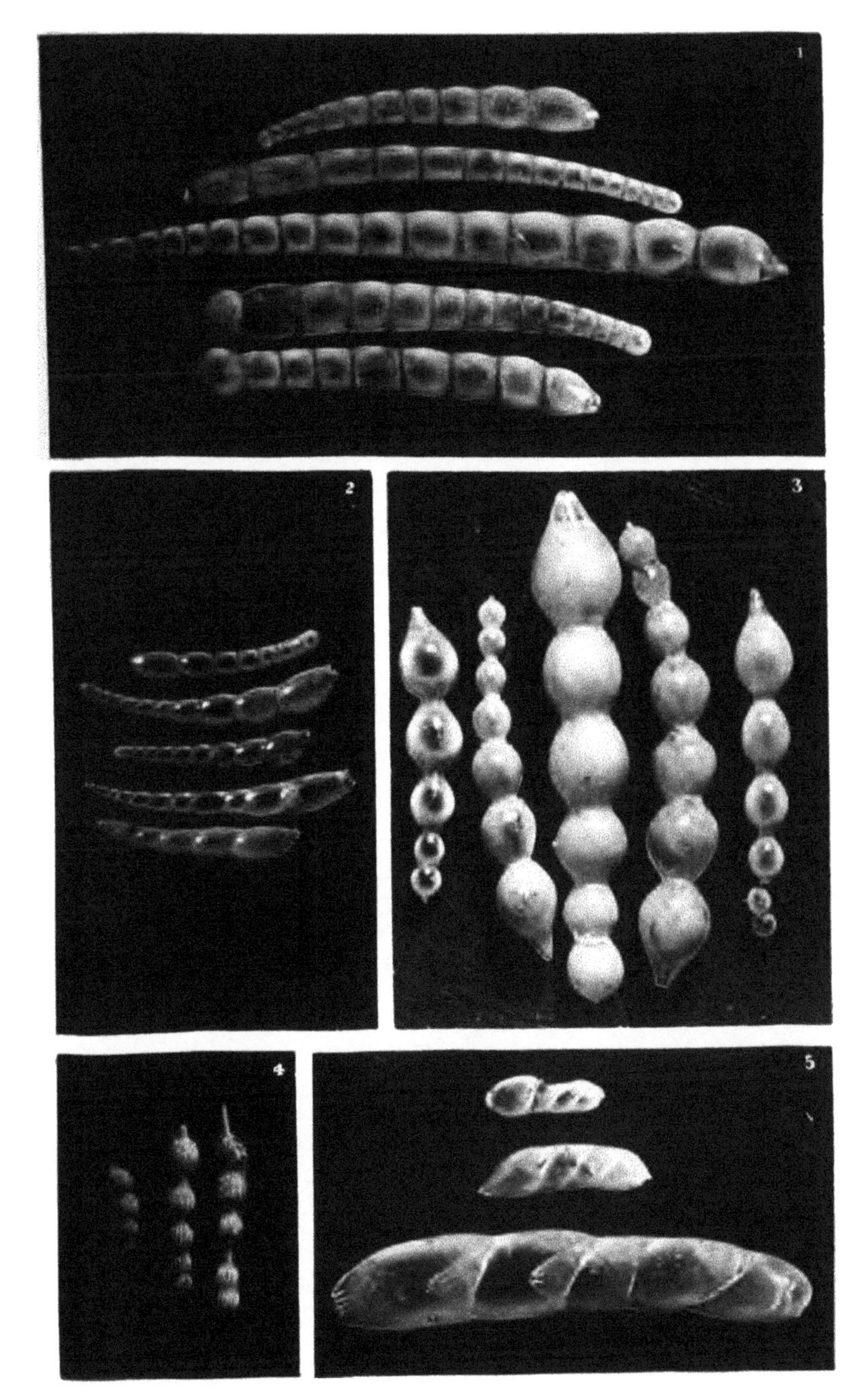

FIG. 1. NODOSARIA CONSOBRINA D'ORBIGNY, VAR. EMACIATA REUSS. SEE PAGE 310

FIG. 2. NODOSARIA COMMUNIS D'ORBIGNY. SEE PAGE 310

FIG. 3. NODOSARIA SOLUTA BORNEMANN. SEE PAGE 310.

FIG. 4. NODOSARIA HISPIDA D'ORBIGNY, VAR. SUBLINEATA BRADY. SEE PAGE 311

FIG. 5. NODOSARIA ROEMERI NEUGEBOREN. SEE PAGE 310.

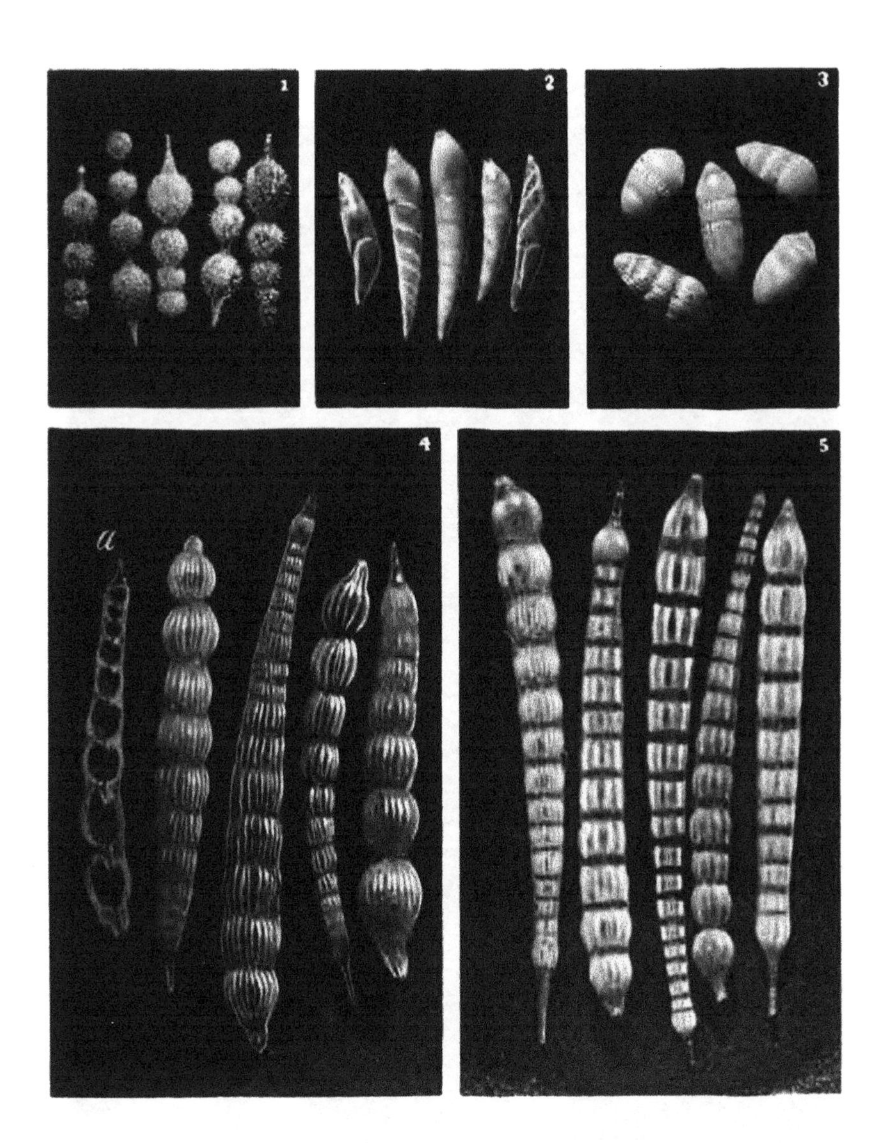

FIG. 1. NODOSARIA HISPIDA D'ORBIGNY. SEE PAGE 311
FIG. 2. NODOSARIA MUCRONATA NEUGEBOREN. SEE PAGE 311
FIG. 3. NODOSARIA COMATA BATSCH. SEE PAGE 311
FIG. 4. NODOSARIA OBLIQUA LINNÆUS. SEE PAGE 311.
a Longitudinal Section
FIG. 5. NODOSARIA VERTEBRALIS BATSCH. SEE PAGE 312.

FIG. 1. NODOSARIA COSTULATA REUSS. SEE PAGE 312.
FIG. 2. NODOSARIA CATENULATA BRADY. SEE PAGE 312.
FIG. 3. LINGULINA CARINATA D'ORBIGNY. SEE PAGE 312.
FIG. 4. LINGULINA CARINATA D'ORBIGNY, VAR. SEMINUDA HANTKEN. SEE PAGE 312.
a. Longitudinal Section

FIG. 1. FRONDICULARIA ALATA D'ORBIGNY. SEE PAGE 313.
FIG. 2. FRONDICULARIA INÆQUALIS COSTA. SEE PAGE 313.
FIG. 3. MARGINULINA ENSIS REUSS. SEE PAGE 314.

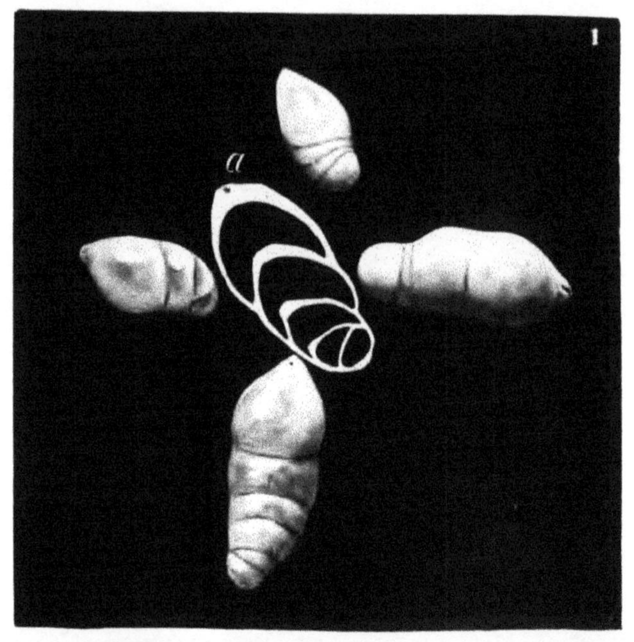

FIG. 1. MARGINULINA GLABRA D'ORBIGNY. SEE PAGE 313.
a Longitudinal Section
FIG. 2. VAGINULINA LEGUMEN LINNÆUS. SEE PAGE 314.
FIG. 3. VAGINULINA SPINIGERA BRADY. SEE PAGE 314.

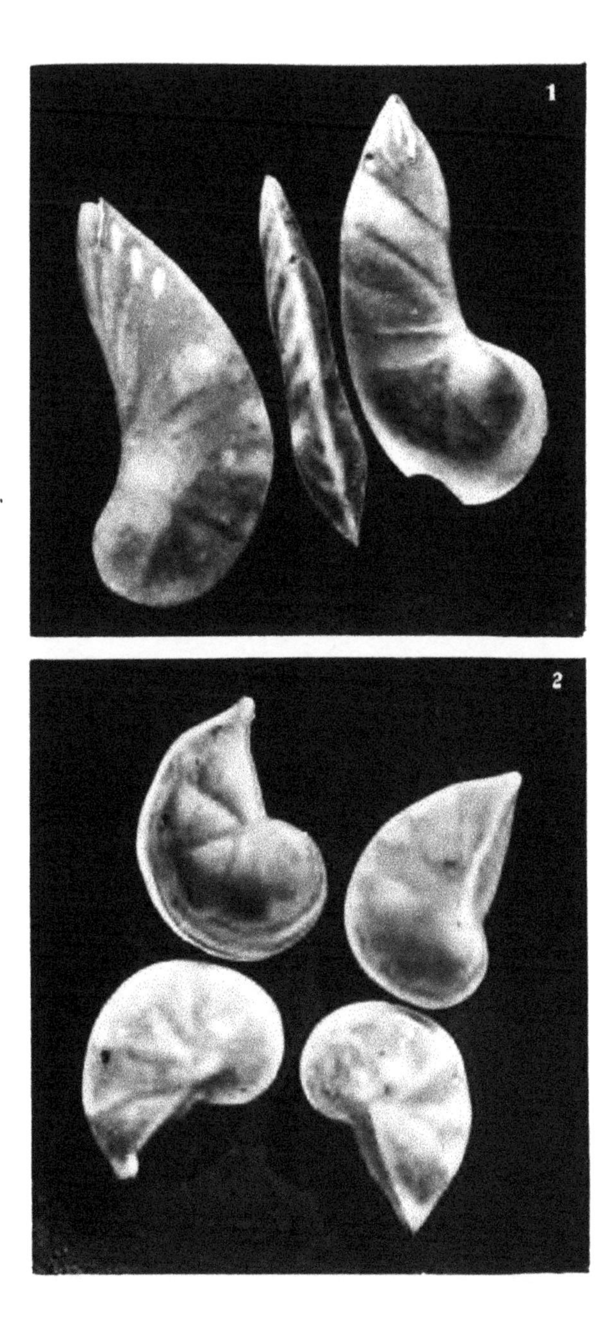

FIG. 1. CRISTELLARIA COMPRESSA D'ORBIGNY. SEE PAGE 315.
FIG. 2. CRISTELLARIA RENIFORMIS D'ORBIGNY. SEE PAGE 315.

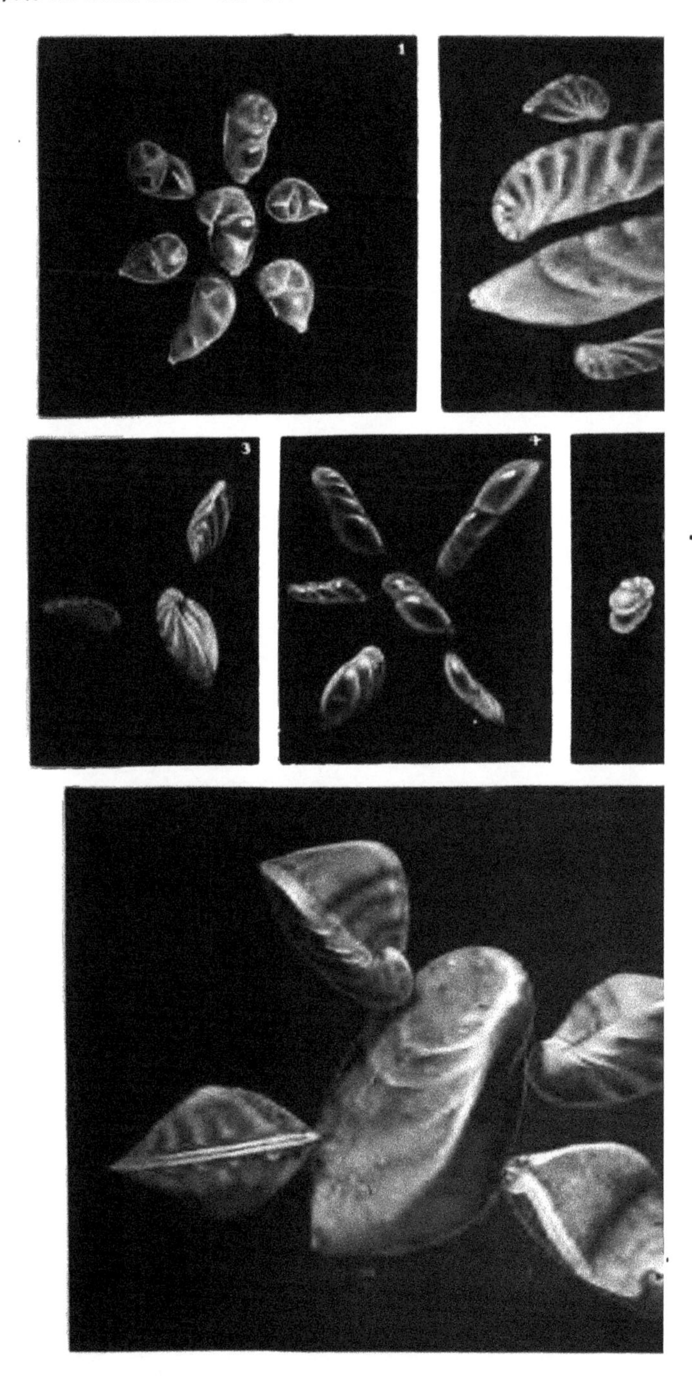

FIG. 1. CRISTELLARIA VARIABILIS REUSS. SEE PAGE 316.
FIG. 2. CRISTELLARIA CREPIDULA FICHTEL & MOLL. SEE PAGE 316.
FIG. 3. CRISTELLARIA LATIFRONS BRADY. SEE PAGE 316.
FIG. 4. CRISTELLARIA SCHLOENBACHI REUSS. SEE PAGE 315.
FIG. 5. CRISTELLARIA ACUTAURICULARIS FICHTEL & MOLL. SEE PAG
FIG. 6. CRISTELLARIA ITALICA DEFRANCE. SEE PAGE 316.

FIG. 1. CRISTELLARIA GIBBA D'ORBIGNY. [illegible]

FIG. 2. CRISTELLARIA ARTICULATA REUSS. SEE PAGE 3[illegible].

FIG. 3. CRISTELLARIA ORBICULARIS D'ORBIGNY. SEE PAGE [illegible].

FIG. 4. CRISTELLARIA ROTULATA LAMARCK. SEE PAGE [illegible].

[illegible]

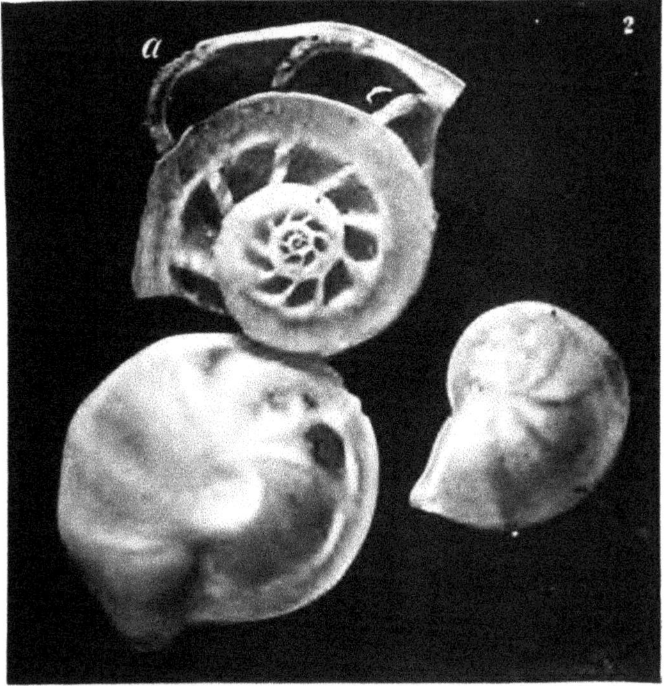

FIG. 1. CRISTELLARIA VORTEX FICHTEL & MOLL. SEE PAGE 317.
FIG. 2. CRISTELLARIA CULTRATA MONTFORT SEE PAGE 318.
a. Horizontal Section.

FIG. 1. CRISTELLARIA CALCAR LINNÆUS. SEE PAGE 318.
FIG. 2. CRISTELLARIA ECHINATA D'ORBIGNY. SEE PAGE 318.
FIG. 3. CRISTELLARIA ACULEATA D'ORBIGNY. SEE PAGE 318.

FIG. 1. CRISTELLARIA LIMBATA NEW SPECIES. SEE PAGE 318.
FIG. 2. POLYMORPHINA SORARIA REUSS. SEE PAGE 319.
FIG. 3. POLYMORPHINA COMPRESSA D'ORBIGNY. SEE PAGE 319.
a Section
FIG. 4. POLYMORPHINA ELEGANTISSIMA PARKER & JONES. SEE PAGE 319.
FIG. 5. POLYMORPHINA OBLONGA D'ORBIGNY. SEE PAGE 319.
FIG. 6. POLYMORPHINA COMMUNIS D'ORBIGNY. SEE PAGE 319.

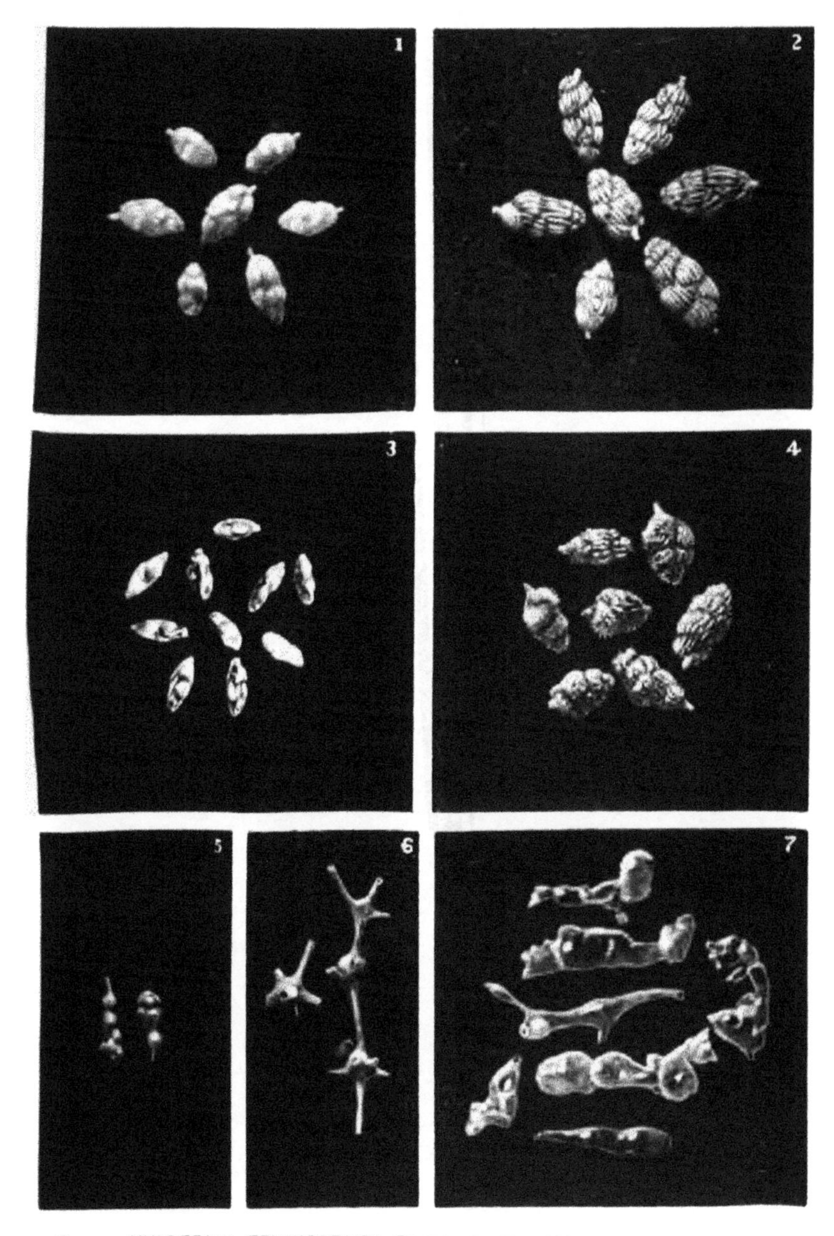

FIG. 1. UVIGERINA TENUISTRIATA REUSS. SEE PAGE 320.
FIG. 2. UVIGERINA PYGMÆA D'ORBIGNY. SEE PAGE 320.
FIG. 3. UVIGERINA ANGULOSA WILLIAMSON. SEE PAGE 320.
FIG. 4. UVIGERINA ASPERULA CZJZEK. SEE PAGE 320.
FIG. 5. UVIGERINA ASPERULA CZJZEK, VAR. AMPULLACEA BRADY. SEE PAGE 320.
FIG. 6. RAMULINA GLOBULIFERA BRADY. SEE PAGE 321.
FIG. 7. RAMULINA PROTEIFORMIS NEW SPECIES. SEE PAGE 321.

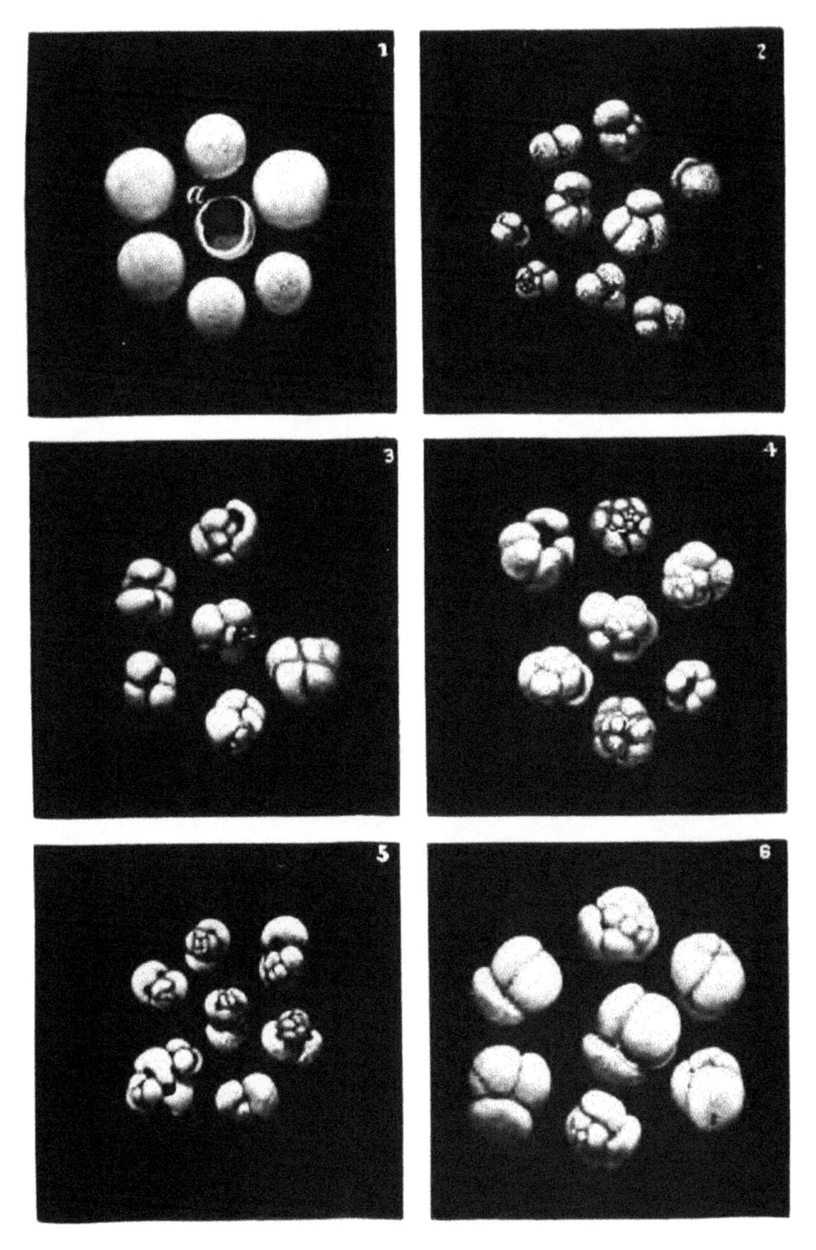

FIG. 1. ORBULINA UNIVERSA D'ORBIGNY. SEE PAGE 323.
a. Accidental Section

FIG. 2. GLOBIGERINA BULLOIDES D'ORBIGNY. SEE PAGE 321.

FIG. 3. GLOBIGERINA INFLATA D'ORBIGNY. SEE PAGE 322.

FIG. 4. GLOBIGERINA DUBIA EGGER. SEE PAGE 322.

FIG. 5. GLOBIGERINA RUBRA D'ORBIGNY. SEE PAGE 322.

FIG. 6. GLOBIGERINA CONGLOBATA BRADY. SEE PAGE 322.

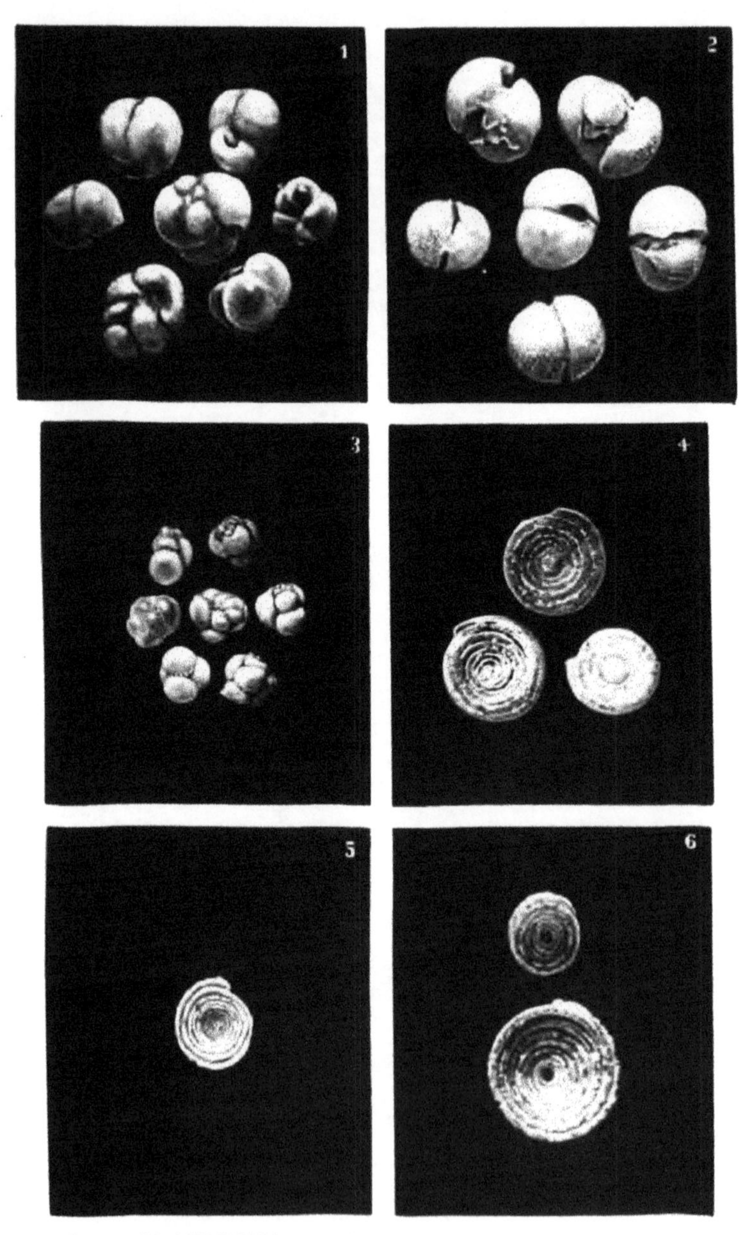

FIG. 1. SPHÆROIDINA BULLOIDES D'ORBIGNY. SEE PAGE 325.
FIG. 2. SPHÆROIDINA DEHISCENS PARKER & JONES. SEE PAGE 325.
FIG. 3. CANDEINA NITIDA D'ORBIGNY. SEE PAGE 325.
FIG. 4. SPIRILLINA VIVIPARA EHRENBERG. SEE PAGE 326.
FIG. 5. SPIRILLINA LIMBATA BRADY. SEE PAGE 326.
FIG. 6. SPIRILLINA OBCONICA BRADY. SEE PAGE 326.

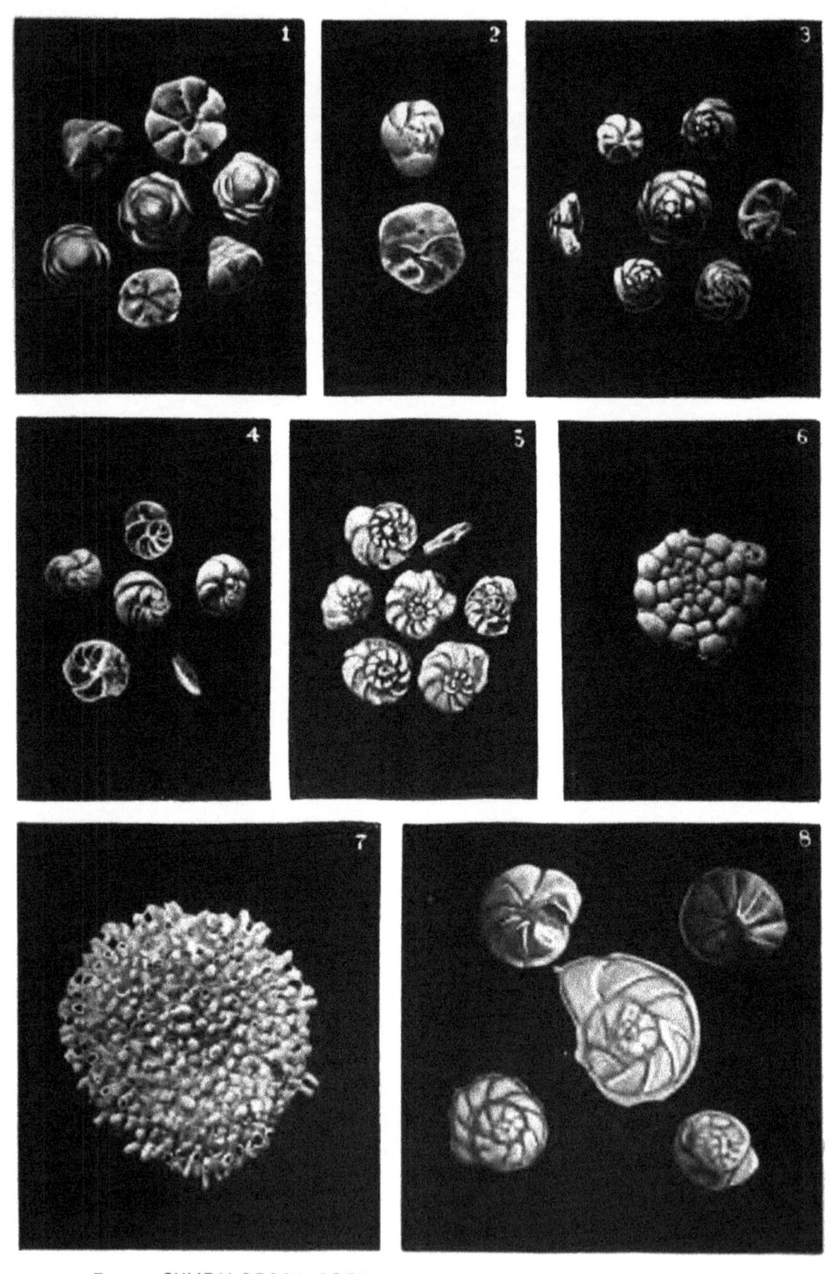

FIG. 1. CYMBALOPORA POEYI D'ORBIGNY. SEE PAGE 326.
FIG. 2. DISCORBINA GLOBULARIS KARRER. SEE PAGE 327.
FIG. 3. DISCORBINA ROSACEA D'ORBIGNY. SEE PAGE 327.
FIG. 4. DISCORBINA BERTHELOTI D'ORBIGNY. SEE PAGE 327.
FIG. 5. DISCORBINA BICONCAVA JONES & PARKER. SEE PAGE 327.
FIG. 6. PLANORBULINA MEDITERRANENSIS D'ORBIGNY. SEE PAGE 328.
FIG. 7. PLANORBULINA ACERVALIS BRADY. SEE PAGE 328.
FIG. 8. PULVINULINA REPANDA FICHTEL & MOLL. SEE PAGE 328.

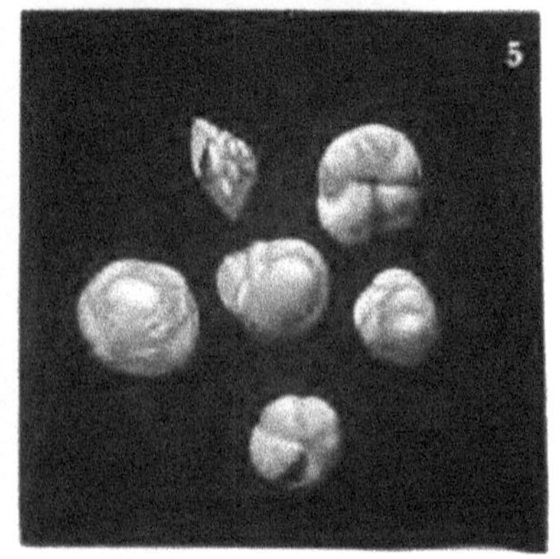

FIG. 1. PULVINULINA CRASSA D'ORBIGNY. SEE PAGE 329.

a. Transverse Section.

FIG. 2. PULVINULINA MICHELIANA D'ORBIGNY. SEE PAGE 330.

a. Partial Section.

FIG. 3. PULVINULINA PAUPERATA PARKER & JONES. SEE PAGE 330.

FIG. 4. PULVINULINA UMBONATA REUSS. SEE PAGE 330.

FIG. 5. PULVINULINA KARSTENI REUSS. SEE PAGE 330.

FIG. 1. ROTALIA SCHROETERIANA PARKER & JONES. SEE PAGE 332.
a. [illegible]
FIG. 2. ROTALIA PAPILLOSA D'ORBIGNY. SEE PAGE 332.
FIG. 3. ROTALIA PULCHELLA D'ORBIGNY. SEE PAGE 332.
FIG. 4. TRUNCATULINA LOBATULA WALKER & JACOB. SEE PAGE 333.

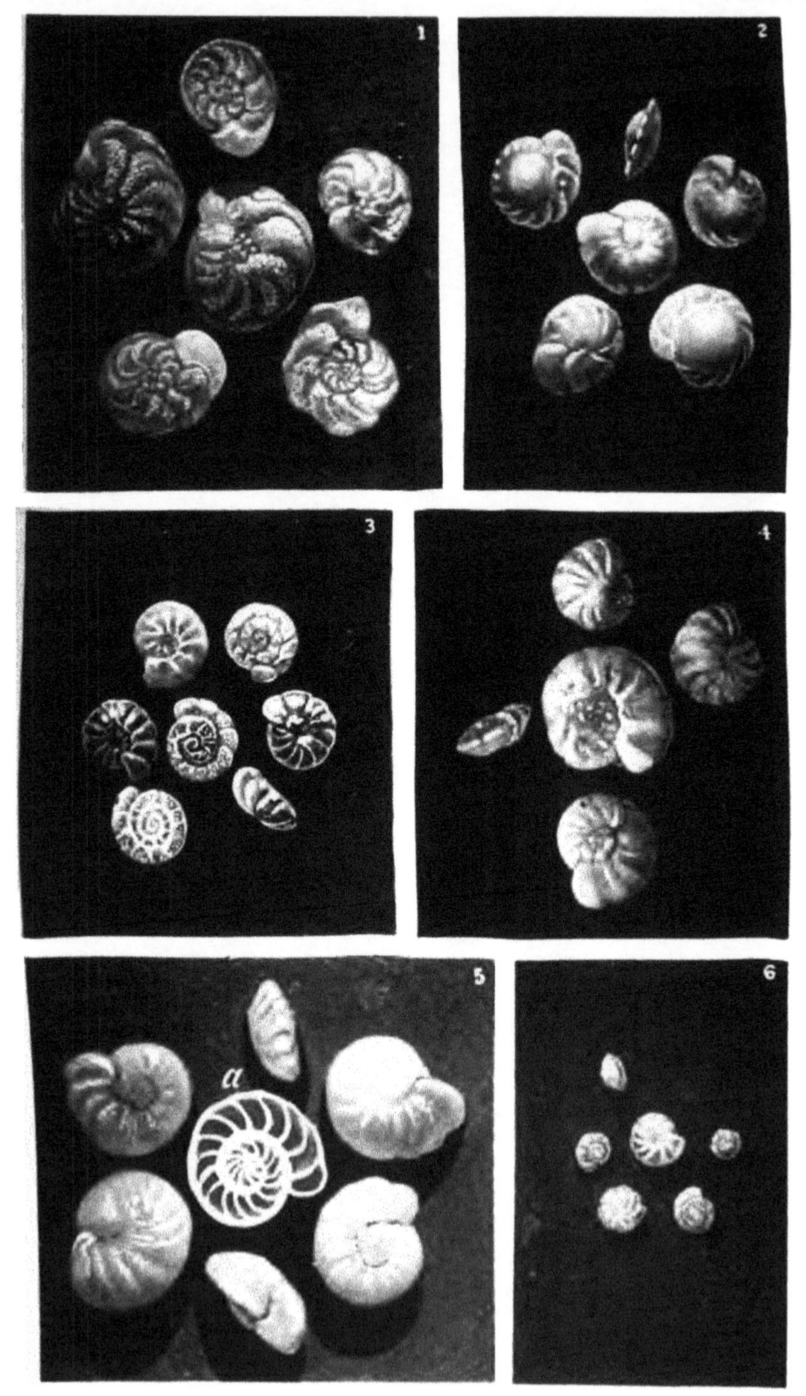

FIG. 1. TRUNCATULINA WUELLERSTORFI SCHWAGER. SEE PAGE 333.
FIG. 2. TRUNCATULINA UNGERIANA D'ORBIGNY. SEE PAGE 333.
FIG. 3. TRUNCATULINA ROBERTSONIANA BRADY. SEE PAGE 333.
FIG. 4. TRUNCATULINA TENERA BRADY. SEE PAGE 334
FIG. 5. TRUNCATULINA AKNERIANA D'ORBIGNY. SEE PAGE 333.
a. Horizontal Section.
FIG. 6. TRUNCATULINA PYGMÆA HANTKEN. SEE PAGE 334.

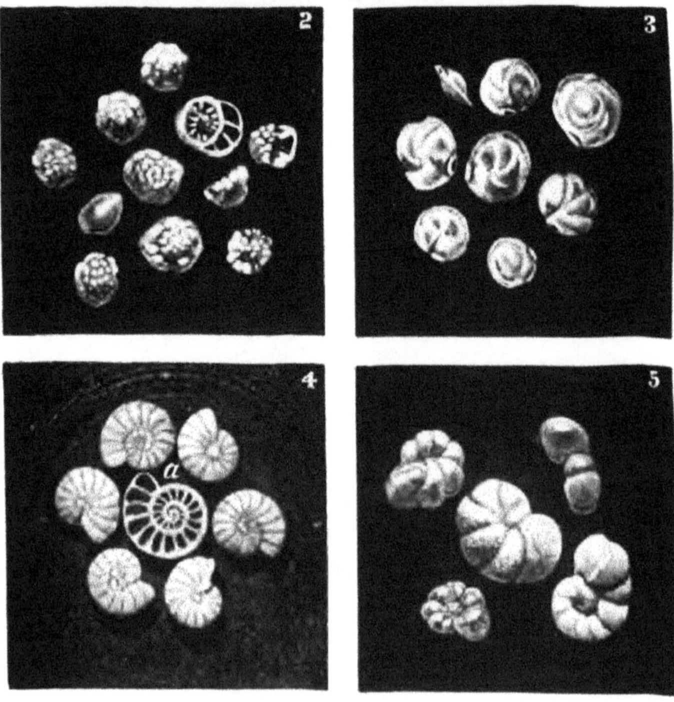

FIG. 1. TRUNCATULINA PRÆCINCTA KARRER. SEE PAGE 334.

FIG. 2. TRUNCATULINA ROSEA D'ORBIGNY. SEE PAGE 334.

FIG. 3. TRUNCATULINA RETICULATA CZJZEK. SEE PAGE 334.

FIG. 4. ANOMALINA AMMONOIDES REUSS. SEE PAGE 335.

a. Horizontal Section

FIG. 5. ANOMALINA GROSSERUGOSA GÜMBEL. SEE PAGE 335.

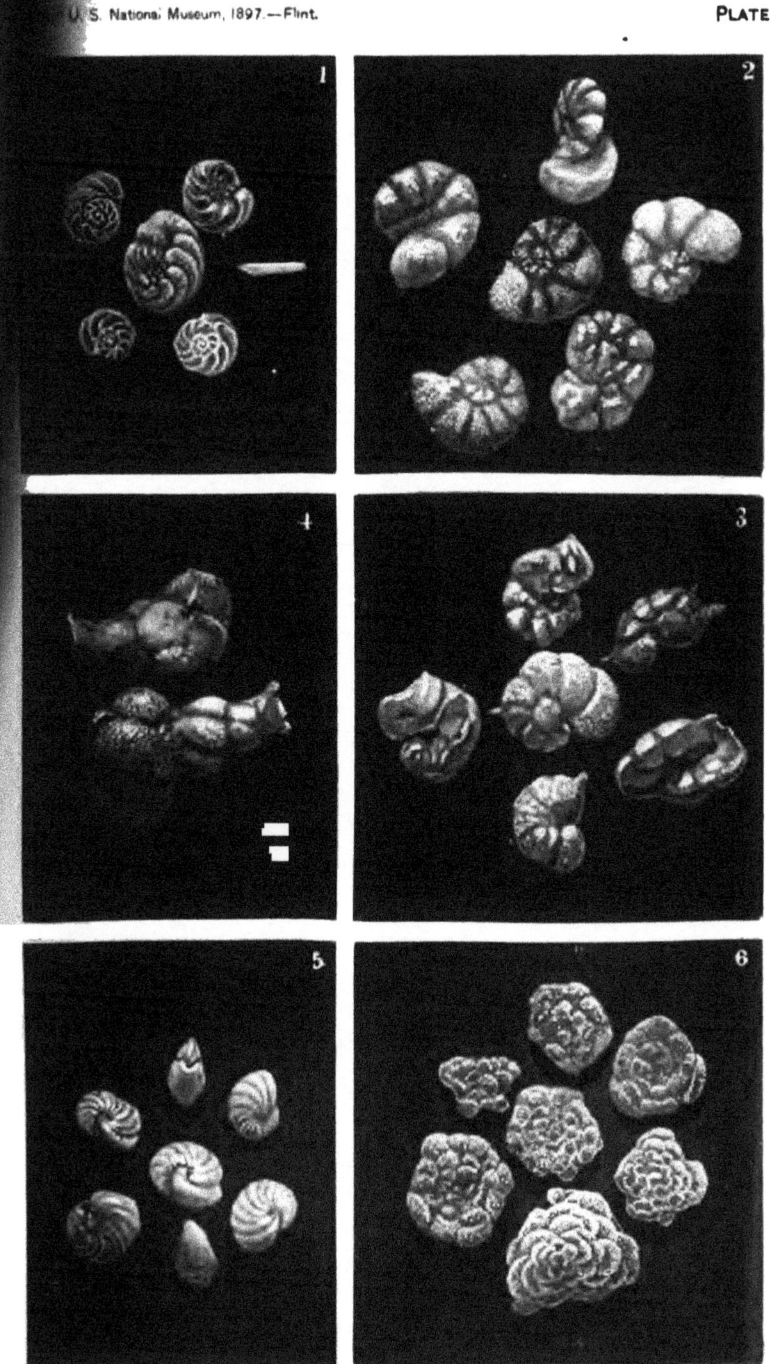

FIG. 1. ANOMALINA ARIMINENSIS D'ORBIGNY. SEE PAGE 335.
FIG. 2. ANOMALINA CORONATA PARKER & JONES SEE PAGE 335.
FIG. 3. ANOMALINA POLYMORPHA COSTA. SEE PAGE 336.
FIG. 4. RUPERTIA STABILIS WALLICH. SEE PAGE 336.
FIG. 5. NONIONINA BOUEANA D'ORBIGNY. SEE PAGE 337.
FIG. 6. GYPSINA INHAERENS SCHULTZE. SEE PAGE 336.

Lightning Source UK Ltd.
Milton Keynes UK
UKHW020704050119
334854UK00004B/402/P

9 780332 681955